The Future of Work

The Future of Work

A Guide to a Changing Society

CHARLES HANDY

Basil Blackwell

© Charles Handy, 1984

First published in 1984 by Basil Blackwell Publisher Limited,
108 Cowley Road, Oxford OX4 1JF.
Reprinted in 1984, 1985

Basil Blackwell Inc., 432 Park Avenue South, Suite 1505,
New York, NY10016, USA

British Library Cataloguing in Publication Data
Handy, Charles B.
 The future of work.
 1. Labor and laboring classes—Great
 Britain 2. Work
 I. Title
 306'.36'0941 HD8391
 ISBN 0-631-14277-0
 ISBN 0-631-14278-9 pbk

Typeset by Katerprint Co Ltd, Oxford
Printed and bound in Great Britain
by Billing and Sons Ltd, Worcester

Contents

Acknowledgements

This book had its origins in the series of consultations that took place in St George's House, Windsor, starting in 1978 when a phrase like 'the future of work' still sounded odd. I am grateful to the many individuals, from all sectors of society, who took part in those consultations and to my colleagues on the staff of the House for all that they contributed to my early thinking on the subject.

My principal intellectual debts are to those authors listed as Further Reading at the end of the book. There are, of course, many more whose thoughts and words have influenced me. I hope I have acknowledged them, where appropriate, in the references for each chapter, but many of the concepts were found in seminars with groups of managers, voluntary workers, politicians, union leaders and others. Their arguments, and in particular their caveats to my propositions, often had a lot more influence than they were aware of at the time.

Sue Corbett helped greatly to shape the book and Elizabeth Bland to polish it. I owe much to them both, and to all at my very efficient publishers. Elizabeth, my wife, typed and typed again the manuscript and supported me in the writing of it. Writing would be a very lonely and empty form of homework without her presence. If the book helps in any way to make the world a better place for our children, Kate and Scott, and their contemporaries to grow up in, then we shall be well content.

C.H.
Diss, Norfolk
March 1984

Introduction

In the late 1970s the elm trees of Britain fell victim to a crippling disease. Although everyone hoped that his or her little patch or prospect would be unscathed, the time inevitably came when the leaves withered on the branches and the next year the bare dead elms stood there, stark skeletons against the sky. When they were felled a part of Britain's ancestry went with them. It was a time for mourning and remembrance. But with their going new vistas opened up and new trees were planted in new formations. Soon it was hard to remember how the church used to look when surrounded by its elms, or the lane when lined with them. Life goes on and a generation is growing up that never knew the elms or its avenues, and will not miss them.

It was during the 1970s, too, that the familiar scenery of our working lives began to show visible changes. The large employment organizations which had been day-time houses for so many people all their lives began to decline. Some famous names from our industrial past disappeared for ever. The names on the High-Street shop fronts changed, and the way of life behind many of them changed too. The tradition of a man going out to work to support, by himself, a family at home became a statistical rarity; by the end of the decade only 14 per cent of households fitted this stereotype. 'Long-term unemployment', 'youth unemployment' and 'redundancy' became familiar words, words which increasingly infected all social groups. Jobs began to be a scarce commodity, and 'work' started to mean other things besides the conventional full-time job. Second and third careers, moonlighting and the black economy became part of our language as did the chip and the video – all new words to herald new ways. The old patterns were breaking down; new patterns were forming.

It will be the argument of this book that there can be no looking back. The new patterns of work are on their way whether we welcome them or not. By the early 1980s the direction of some of those patterns of work was becoming clearer:

(1)   the full-employment society was becoming the part-employment society;

(2)   'labour' and 'manual skills' were yielding to 'knowledge' as the basis for new businesses and new work;

(3)   'industry' was declining and 'services' were growing in importance;

(4)   'hierarchies' and 'bureaucracies' were going out, 'networks' and 'partnerships' were coming in;

(5)   the one-organization career was becoming rarer, job-mobility and career changes more fashionable;

(6)   the 'third-age' of life, beyond the ages of growing-up and of employment, was becoming important to more and more people;

(7)   sexual stereotypes were being challenged, at work and in the home, and roles were no longer rigid;

(8)   work was shifting southwards, inside countries and between countries.

Taken separately, each of these changing patterns could perhaps be absorbed without too much trouble. Together they add up to a frontal assault on our view of work, the way in which it is organized and the place it occupies both in our lives and in the way that society is run. The consequences will be painful and demoralizing to many – that seems inevitable – and it is only natural that there will be many who will say that it need not and should not happen. On the other hand, if we can look beyond what is being lost to what will replace it, there are these glimpses of new patterns which offer their own excitement if we are prepared to see them that way, and if we are willing to equip ourselves and others to take advantage of them.

Indeed, to our grandchildren the massive organizations of this industrial age may look as bizarre as trench warfare does to today's military commanders. The idea of the 100,000 hours, 47 hours per week for 47 weeks a year for 47 years that everyone used to work, and many still do, may seem as unnatural to them as child labour in the mines does to us, the image of a society built on factories and industrial towns as remote, and perhaps as nostalgic, as the old rural villages of the agricultural age.

Meantime, what we recognize is a confusing set of paradoxes. First, society appears to promise what it cannot deliver. All Governments commit themselves to full employment but in the past ten years every Government has had to watch the unemployment figures rise. Second, while people clamour for work a lot of work remains undone. Skill shortages are widespread in certain types of work and vacancies remain unfilled around the country. Streets could be cleaned better, buses scheduled more frequently, children taught more intensively, old people cared for better. Third, we want work but do not always like it. Many find employment degrading and unpleasant. Six million people change their jobs each year, many because they are dissatisfied. Fourth, we say we only work to live, but we cannot live without work. Leisure without work to set against it is unappealing to many. Life without work leaves them without purpose or identity.

These paradoxes have emerged, I will argue, because the employment society has overstretched itself. We have, to put it crudely, made the job the only legitimate form of work and then priced many jobs so high that we have effectively priced them out of existence. The answer is not to make a universal and drastic cut in wages, which would be both impractical and unfair, but to look beyond employment to a society which is less job-fixated in its values, its structures and its systems. The signs are that we are already being forced into a world where the job is a necessary part of life but only a part of it. Jobs for all, yes, but no longer jobs for all for always.

It is the central argument of this book that things *are* changing, that work will not be the same again and that it is only prudent to consider what it will be like and what needs to be done, now, to make sure that we make the most of the new possibilities and that the transition from the old to the new is not too painful. We need to plan and to prepare. If we do not, the pain and the problems will obscure the opportunities. The new scene could then be a bleak one for most of us. The book is intended as an alarm bell to waken us up as well as an invitation to join a potentially exciting kind of society.

Is it, however, going to be *that* different? To what extent *are* things changing? There are many who see our present difficulties as but a stage in history's self-correcting cycle of adjustment; we only have to endure, they argue, and the old patterns will come into their own again. They look to the following precedents:

(1) *We have created jobs before.* Between 1932 and 1937 we increased the number of jobs by 2.25 million and reduced unemployment by 1.25 million.

(2) *We have lived with a growing labour force before.* In the 100 years from 1860 to 1960 the labour force nearly doubled, but so did employment.

(3) *We have survived structural change before.* Agricultural jobs fell

from over 25 per cent of the labour force to under 3 per cent in the same 100 years but were all absorbed into new industries and occupations.

(4) *We have seen technological change before.* Between 1860 and 1960 the capital employed per worker doubled, but output trebled. Automobiles replaced horses and society adjusted.

That society will, in time, adapt is certain. Is it equally certain that it will always adapt in the same way?

The argument in this book starts from the assumption that what we are experiencing is more than a cyclical adjustment. We are seeing the gradual tapering-off of the employment society, a society in which jobs are the measure and the means of most things. An employment society generates its wealth through jobs, the more people working in an organized formal way the greater the transferable wealth, as one man's earnings from his job allows him to purchase another man's output. In an employment society the job is the way in which society distributes wealth among its citizens, it is the financial and the social lynchpin of most people's lives and therefore, naturally, the source of much of the meaning of their existence. Take away jobs in an employment society and you not only take away the means of access to many of the pleasures and opportunities of modern society, but you also call in question the whole point and purpose of life without a job.

If the employment society is past its peak then we shall have to re-think many of those assumptions with which we have lived so comfortably for over three hundred years. If jobs cannot be taken as given, how will people get money? What will they do with their time? How will they measure wealth, or success or happiness? What structures will they find to give pattern to their existence? A changing future of work begs some of mankind's most fundamental questions. It is not a purely technical matter nor a thing of interest only to economists and their friends. It is the stuff of life itself.

It is therefore crucial to face up to the question of whether things are changing – and, if so, how – before going on to look at the implications of any change. Chapter 1 is devoted to a discussion of this issue. Chapter 2 looks in more detail at what is happening to jobs, only to conclude that the answer cannot lie in more conventional work. Chapter 3 then examines the options beyond employment if we broaden out the meaning of work, to include marginal work and gift work as well as the job. Chapter 4 explores what happens to organizations and individuals as jobs get whittled away and rearranged with the help of new technology, discussing the emergence of the 50,000 hour job, the contractual organization and the federal structure. Chapter 5 looks at the social and political agenda that is opened up by the new patterns of work, from taxation and wealth creation, to housing and social welfare, leaving chapter 6 to deal with the crucial question of education and the ways in which it needs to change and develop. Chapter 7 seeks to focus the issues by looking at the choices that have to be made by individuals in planning their lives and choosing their families and by society in balancing the conflicting requirements of liberty and equality, and the different forms of justice.

Inevitably and deliberately, this sort of book is a personal view. It is not intended to be a comprehensive review of the growing futurist literature. It is not an essay in economics, or in management, or even in individual psychology, although all these subjects come into it. It is an attempt to stimulate the thoughtful person at home or in the office to think about the way in which the world of work is changing and what his or her choices and responsibilities should be.

A book as short as this on a topic as broad as this has to dwell only fleetingly on areas of social, fiscal and political argument which are highly complex, technical and contentious, but readers who wish to delve deeper should find enough guides listed in the references for each chapter.

The one certain thing about any view of the future is that it will be wrong in many respects. The importance of the arguments in this book lies not in any claim that they are right but in the suggestion that they are plausible enough to arouse discussion and counter-argument. The future is too important to be left to chance.

1

A Changing World?

This chapter addresses three questions:

Will there be enough jobs in the future?
What kinds of job will they be?
What else will we do?

The answers point to a new agenda for all concerned with the future of work and of our society. Because there is little point in reading the rest of this book if you do not believe that things are changing, these questions will be examined briefly here before we return to them more fully in subsequent chapters.

WILL THERE BE ENOUGH JOBS IN THE FUTURE?

The signs are, to put it bluntly, that there are not going to be enough conventional jobs to go around – not full-time, lifetime jobs with an employer who pays you a pension for the ten years of so of your retirement. That is going to be true no matter which Government is in power over the next twenty years.

Chapter 2 will unravel the figures for Britain in more detail, but in essence jobs have been leaving industry and business for the state sector for over twenty years. Until 1980 the state managed to create as many jobs as business lost, but in 1980 that stopped. In the meantime the labour force had grown over the past twenty years by 2.5 million, mainly because more married women wanted jobs. The potential workforce is likely to grow by another million over the next fifteen years because the babies born in the 1960s, who are now growing up, outnumber their parents, who might be retiring. In other words, just when the labour force is growing, employment is falling. The result is an increasing gap – unemployment.

The continuity advocates claim that the gap can be closed by creating more jobs. One school wants an expansion of the business sector, particularly in manufacturing. This is to look for a reversal of the trend of the last twenty years. It assumes that an increase in output necessarily means an increase in jobs, whereas all the signs are that we are moving from labour-intensive businesses to labour-scarce ones. The reasons are obvious. If labour is relatively expensive, you need businesses and processes which are capital- not labour-intensive. The only business sectors to create more employment in the past decade were banking and insurance.

Yet these two sectors, which were both labour-intensive, are rapidly becoming capital-intensive as a result of the new information technologies. Businesses need to improve their productivity continually just to stay in business. Their output must grow by at least as much as that productivity improvement if they are to keep the same number of jobs. A productivity improvement of 10 per cent, given the same quantity of output, is a polite way of saying that the labour force has been decreased by 10 per cent. Output must, in fact, grow faster than productivity if a business wants to pay higher wages to the same labour force and it must grow faster still if it wants actually to increase the number of jobs. This is the 'productivity trap'. It applies both to individual businesses and to the business sector as a whole. We may look to business for more output – indeed we must – but not for more jobs unless we think we can grow much faster than our competitors or abandon the need to compete on efficiency (i.e. move to a protected economy and try to survive on what we can make for ourselves, like some Eastern bloc countries). Extra high growth or a siege economy: these are the two ways of avoiding a declining industrial workforce. The first is implausible, the second undesirable.

Another school of thought wants government to close the gap, by being a more adventurous customer for infrastructure works or by employing more people itself, particularly in health and education, which have been the two growth sectors for jobs in the past decade. There are, however, two problems – the numbers and the cost (see chapter 2).

The likelihood is that during the next two decades we shall see some of each of these initiatives in attempts to close the gap. But the most optimistic assumption is that the creation of new conventional jobs in business and government will be hard-pressed to keep pace with the growth in the labour force, leaving the unemployment gap varying between 3 to 4 million. It could be much bigger if industry continues to slim down its workforce and if government fails to take up the slack. *There will not, therefore, be enough jobs.* That much seems certain. (The figures will be looked at more closely in Chapter 2.)

We could, at this point, stand back and let employers enjoy the resulting buyers' market for jobs and then find some way of supporting the rest in their involuntary leisure. That might be justified in the name of economic efficiency but hardly in the name of social justice.

If we cannot do all that much to increase the *supply* of jobs, then perhaps we can close the gap by reducing the *demand* for them. People could be kept longer in education (by the Youth Training Scheme, for instance) or encouraged to retire early. Some have even advocated that households should be rationed to one job each. Alternatively, the available jobs could be split up by various schemes of work-sharing so that more people had at least part of a job. Contrary to popular impression, part-time work appears to be increasing in Britain as does self-employment, which is as often part-time as overtime.

In other words, if we cannot give everyone a job for the whole of their lives would it be preferable to give everyone a job for part of their lives rather than give it to some for all of their lives and to some for none? What would this mean in practice? It might be that we started work late and ended it early in our lives. Or would it mean more sabbaticals and longer holidays, or shorter weeks and more part-time work? If so, what would happen to our earnings? If we could maintain the same productivity with fewer years or fewer hours, there would be no need for new jobs. If we could not, would we be prepared to accept less money for less work? These are critical questions, but they will come to the front only when it is finally, if reluctantly, accepted that the job gap will not and cannot be closed totally by an increase in the supply of jobs; moreover, it has first to be accepted that there is no way out of the dilemma if a job continues to mean what it means today.

This is not just Britain's problem, it is the world's. The International Labour Organization (ILO) estimates that 1,000 million new jobs would have to be created between now and the year 2000 to achieve full employment worldwide. But, says the Director-General of ILO 'It has to be fully understood that there will be no situation of full employment if we are speaking of *conventional* employment . . . I am convinced that over the next twenty years we shall see a change in the nature of employment.'

This book does not condone unemployment, ever, for anyone. It is not a paean for an age of leisure, forced or voluntary. It is a worried and worrying exploration of the kind of world we are entering where the supply of jobs, as defined today, is unlikely to be adequate to meet the demand. It will suggest that while every effort should be made to increase output, both nationally and internationally, this can never be enough to solve the problem of

BOX 1.1   ECONOMISTS AND UNEMPLOYMENT

Economic thinking on employment and unemployment has always been dominated by two thoughts: 'Work is always available for those who want it' (enshrined in the Statute of Artificers in 1563), and the later caveat 'provided they accept the going price for their labour'. The argument has really been about the 'frictional resistances' to the full working of those two principles.

Say's Law, which states that supply creates its own demand – that if you produce something, that act generates income, which then demands more production – has not really been disputed, just amended in detail. Even Keynes, who said he was attacking it, was really explaining that it does not always work automatically, particularly in a recession, because of the psychology of consumers and entrepreneurs who are both liable to hold back, especially if encouraged to do so by high interest rates; therefore the Government has to intervene to create a positive climate by starting the spending cycle.

Marx too, did not so much dispute the two thoughts as explain that the going price for labour was always in conflict with the price of capital which he saw as falling and winning and therefore ultimately as always leading to a growth in the 'reserve army of the unemployed'.

Monetarists believe that too much money in the system drives up the price of labour which in the end makes the product too expensive to sell; therefore they wish to ration the money in the system. In practice, however this has meant raising the cost of money and cutting back government spending, creating the very conditions Keynes was anxious to avoid.

It is hard for the dispassionate observer not to feel that they have all some of the truth on their side. The problem is that their discussions draw the boundary too tightly around the word 'work' in the original thoughts. The two starting thoughts are correct: work is always available to those who want it if work is priced at anything down to zero, or is voluntary. They do not make sense if confined to formal employment because the 'frictional resistances' to lower wages are very real. Wages do not go down in unemployment, they only go up more slowly. Therefore labour costs to an employer are cut not by lowered wage rates but by reduced numbers of jobs. 'Work at some prices is always available' is a true statement. 'Employment at acceptable prices is always available' cannot be true except in occasional and exceptional circumstances. We have to go beyond employment to find the long term answer to the economists' conundrum.[1]

unemployment. Indeed, it may well actually *decrease* the number of jobs available, since improved efficiency is the first route to improved output. We have to look beyond economics to the definition of a job, to the meaning of work and the measure of success and meaning in human life.

In crude terms the full employment equation works only if we can find ways of giving more people a smaller slice of employment and that only works if we redefine employment, work and the job. We must look beyond employment not to unemployment or boundless leisure but to a new view of work, of which the job is only one part. The trend is already moving in that direction.

WHAT KINDS OF JOB WILL THEY BE?

The job count is only part of the problem. We are also witnessing a change in the nature of jobs. Muscle jobs are disappearing; finger and brain jobs are growing or, to put it more formally, labour-based industries have been displaced by skill-based industries, and these in turn will have to be replaced by knowledge-based industries.

The reasons, again, are clear. Labour-based industries go in the end to countries where labour is cheap if the labour cannot be replaced by automated equipment. Even Singapore, with labour costs one-eighth of those in Britain, now regards itself as too expensive for labour-based industries. 'Our aim is modest,' said Mr Goh Thock Tong, Minister for Trade and Industry in Singapore, '[It is] to step into the shoes left behind by countries like Germany and Japan as they restructure – they from skill-intensive to knowledge-intensive industries and we from labour-intensive to skill-intensive industries.' The message for Britain is clear.

Mr Goh is quoted by Professor Tom Stonier in his book *The Wealth of Information*,[2] in which he argues that Britain's commercial future lies in information; in the creation of knowledge, its application and its communication. That definition covers a surprising range of activities from consultancy and financial services, through advanced engineering design, architecture and R. and D., to education, publishing, tourism and the arts – all areas in which Britain has a traditional reputation. Professor Stonier is surely right in this; unless we can top up our skills with new knowledge we shall have no commercial advantage.

That means, in practical terms, that there are going to be more offices than factories. It means more schools, universities and colleges than construction yards; it also means that factories will often have polished floors and white-coated workers, like the pharmaceutical factories today, rather than concrete areas for foundries, steel-making and cars. There will always, of course, be splendid exceptions to a general trend. Specialist shipbuilding, special steels, up-market cars, aerospace projects, tanks and war-ships—all these and other bits of heavy industry which apply advanced know-how will have their place in a knowledge-intensive economy. We will not survive if we make only books and computer programmes.

The knowledge orientation is one aspect of the new world of work. There are two other aspects which are less frequently commented on, but will be equally important. The first is the re-emergence of 'gangs'. Technology will make it both possible and desirable for work once again to be organized around gangs, as it used to be before the assembly line and the bureaucratic organization effectively put people into lines in the factories or offices and on the organization charts. Sophisticated automated or robotic machinery can now equip a group of people to do what required a mini-organization to do even ten years ago. The gang, a technically self-sufficient work group, has always been the preferred British way of working. It was only the technology of the assembly line which imposed on us the micro-division of labour (as it is called), the breaking down of work into its smallest components. The job shop, or project group has been a familiar aspect of knowledge-based firms in consultancy, advertis-ing, merchant banking, architecture and R. and D. because technology made it possible. The growth of knowledge industries and the automation of labour and skill industries make similar organization possible almost everywhere.

It could go farther. The gangs would not need to be under the same roof or even in the same employ. Communications are now so quick and so multifaceted that it would almost be easier to communicate sensibly if the gangs were a hundred miles apart than if they were next door to each other. The dispersed organization is becoming increasingly common, particularly in the information business, such as insurance or banking. But if the gangs were technologically independent, why could they not be organizationally independent – commissioned or contracted to produce given quanti-ties of work? It is the way in which much of Japanese industry is organized. We are familiar with subcontracting in construction. We see it in Channel 4 television. It is practised by Marks and Spencer as a formal policy and is the traditional way of life of publishers, who contract the writing to an author, the production to a printer and the selling to booksellers. As the knowledge side of work expands and gangs proliferate, so the idea of the contractual organization will begin to jostle with the traditional notion of the employment organization. (These ideas will be explored in more depth in chapter 4.)

The third aspect of the new world of work is different again. Although the knowledge industries may be displacing the labour and skill industries, there is good evidence that the bulk of new jobs is going to come not from this source but from something quite different – the personal service sector. A feature of the emerging knowledge-based societies is the proliferation of small businesses which grow up around the affluent parts of them, turning into formal economic activities the things that many people normally do for themselves. For instance, in the United States between 1973 and 1980 three industries – 'eating and drinking places', 'health services' and 'business services' – increased their employment at three times the pace of total private employment and sixteen times faster than employment in the goods-producing sector: 'The *increase* in employment in eating and drinking places since 1973 is greater than *total* employment in the automobile and steel industries combined.'[3] A list of some of the personal service businesses is provided in box 1.2. It may well be that the employment discarded by the industrial sector may be picked up by this personal service sector *if* we can remain affluent enough to afford them. We must note in passing, however, that this sector is made up of self-employed people, partnerships and very small businesses. It is entrepreneurial,

BOX 1.2   SOME PERSONAL SERVICES

Contract domestic work (e.g. cooking, cleaning, garden maintenance, child care, building and maintenance)

Eating and drinking (e.g. restaurants, wine bars, fast-food chains, hotels and motels)

Business services (e.g. accounting, tax advice, courier services, telex and typing bureaux)

Craft work (e.g. handmade domestic objects, furniture, paintings, sculpture)

Education (e.g. music, drama, literature, special tutorials, skill training and retraining)

Tourism (e.g. fairs, exhibition parks, guided trips and special tours)

Health services (e.g. fitness and beauty places, sports, hairdressing, dentistry)

---

personal, moderately but not highly skilled; it uses accessible rather than high technology; it is unisex and not very capital-intensive. It is above all, labour-absorbing rather than labour-saving, for, although efficiency and productivity are important, the personal nature of the work limits the scope for substituting capital for labour.

The personal service businesses will undoubtedly flourish in the affluent parts of the new world of work and they do offer exciting entrepreneurial opportunities, but the jobs which they provide are not the traditional lifetime, secure, well paid jobs of the big employment organization. Work in the personal services sector is often erratic, seasonal, insecure and unprotected by formal agreements. It is a world for the young without too many responsibilities, for the entrepreneur or, sometimes, for the desperate. It is, as they say, an open labour market, where there will be jobs, yes, but not always at the price you might like.

WHAT ELSE WILL WE DO?

Jobs will be fewer and farther between. Jobs will be shorter as the 100,000 hours gets whittled away and we work shorter weeks, years and lives. Jobs will be difficult, more dispersed and, in many cases, more precarious. All in all, most people will end up earning less from their jobs *over their lifetime*, even if their actual rates of pay while they are in a job go up. We shall tend, therefore, to have more time but less money, particularly in late middle age. Does that have to be the bad trade-off that anyone brought up in the employment society automatically assumes it to be? Has the job always been the passport to life and happiness that many would like it to be? Opinions are, and always have been mixed (see box 1.3). More time and less money is a good or bad bargain depending on what use we make of that time and on what we value in life.

What, for instance, do we want from work? Most research places identity, friendship and status, the sense of creating something, of contributing and of achieving, above the need for material reward, although the material reward leaps up higher for those out of a job. But do we need to be dependent on the job for all these gratifications? Do all the eggs of life need to be in the one basket? Many people have shied away from the one-basket approach and have tended to put together a portfolio of activities and relationships, each of which makes its own contribution to the package of things we want out of work and life. Will the portfolio approach become more common as we are forced to be less dependent on fading jobs? If so, do we have enough variety of opportunities to find those other activities and other relationships beyond employment?

BOX 1.3 TWO VIEWS OF EMPLOYMENT

Which is yours? Can they both be true?

Full employment is . . . not merely a means to higher production and faster expansion. It is also an aim in itself, weakening the dominance of men over men, dissolving the master – servant relation. It is the greatest engine for the attainment by all of human dignity and greater equality.

Thomas Balogh
The Irrelevance of Conventional Economics (1982)

'I'm a machine,' says the spot-welder. 'I'm caged,' says the bank teller and echoes the hotel clerk. 'I'm a mule,' says the steel worker. 'A monkey can do what I do,' says the receptionist. 'I'm less than a farm implement,' says the migrant worker. 'I'm an object', says the high-fashion model . . . Nora Watson may have said it most succinctly: 'I think most of us are looking for a calling, not a job. Most of us, like the assembly-line worker, have jobs that are too small for our spirit. Jobs are not big enough for people.'

Studs Terkel, *Working* (1974)

Can we live with less money and be satisfied? More time and more accessible technological aids may allow everyone to be more self-sufficient, all doing for themselves what they had previously paid others to do for them. In the years from 1959 to 1982 the percentage of women who were involved in DIY activities in the home rose from 48 to 60 per cent, while for men it went up from 55 to 81 per cent.[4] But DIY need not be confined to building, decorating and gardening. We can educate ourselves and our children, assemble our own appliances, make our own clothes, set our own hair, look after our aged or sick kin more easily now than in the past. We can, in fact, choose either to purchase all the personal services listed in box 1.2 or do them for ourselves, substituting our time for our money. Would we be worse off?

It all depends on how one measures success. For Governments a thriving domestic self-help economy is not to be desired. Fewer traded services mean less taxable revenue, less control and regulation over safety and less measurable wealth to stimulate the formal economy. But individuals who have made the trade-off between time and money often find that the freedom provided by one's own capability is more lasting than the precarious freedom of money to buy in what is needed. As the trade-off is forced upon more people, will the conventional measures of success be changed, so that independence is given a value alongside that of material wealth?

It will, like most things, probably go both ways. As the knowledge industries flourish along the M4 belt in Britain or in the sun belt in the USA, the personal services will bask in their affluence. In the darker areas of the north or the retirement coasts of the east and south the domestic economy may grow. It is important, I shall argue, for both to be legitimate. Jobs must not be the only path to respectability in a society which can guarantee jobs only for part of our life or, perhaps, to part of our society. The domestic economy need not be relegated to the unemployed, the old and the young as a waste heap for the jobless. It could have its own worth, particularly for those in the extended third age of life, beyond employment. No discussion of the future of work can ignore an economy which takes up half of our productive hours and, arguably, contributes a lot to our quality of life. Work is more than jobs, as chapter 3 contends.

A NEW AGENDA

The signs of the future are with us now. The future is not that far away. Guessing at the future patterns of work is not therefore a fashionable game of social fiction; it is the prelude to a serious consideration of a whole range of political and societal dilemmas.

If the trends and portents turn out to be significant, we are likely to find:

(1)   many more people than at present not working for an organization (what will they be doing? How will they be organized?);

(2)   shorter working lives for many people (what will they do? How will they be paid?);

(3)   fewer mammoth bureaucracies, more federal organizations and more tiny businesses (how will they be managed? To whom will they be accountable?);

(4)   more requirements for specialists and professionals in organizations (how will they be trained and retrained? What will the rest of us do?);

(5)   more importance given to the informal, uncounted economy of the home and the community (will homework be real work? How will it be recognized and rewarded?);

(6)   a manufacturing sector that is smaller in terms of people but larger in terms of output (what will it be making? How will it be fed with new ideas, skills and finance?);

(7) a smaller earning population and a bigger dependent population (how will wealth be distributed? What taxation will be appropriate?);

(8) a greatly increased demand for education (for what? By whom? At what age? How will it be paid for?);

(9) new forms of social organization to complement the employment organization (what will they be? Will the family be one of them? Will they be political?).

There are questions here for politicians, managers, educationalists, financiers, the voluntary world and the unions. We shall elaborate on the questions and some of the options below, but perhaps the most important questions will be for the individual growing up today to work in tomorrow's world. It will not be the world that his or her parents or teachers knew. No longer will the question 'What will I be?' be fully answered by 'What job shall I have?' The job will no longer be the whole measure of one's identity, one's status, one's finances or one's purpose in life. This may be the opportunity for a wider vision of humanity and of life. It is also an opportunity that many would prefer not to have. What will success mean if it cannot be defined in terms of money and job? Will there be enough money anyway, or is happiness going to have to be independent of prosperity? Or will it be that old Catch-22 situation: only when you have money can you afford to despise it? The changing shape of work starts with technology, economics and demography but inevitably ends up in philosophy. It is not a new conclusion. The economist Maynard Keynes, writing in the middle of the 1930s recession, saw it coming.

It is an agenda and a set of scenarios which has more appeal, and apparently more to offer, to the well educated member of the professional middle classes. Those with credentials and saleable skills, who know how to manage their own lives, can sell themselves to others and realize that sacrifice today for rewards in the future is sensible investment. These will prosper in the world of work that is coming up. Ironically, it is the so-called working classes who could find themselves without work in the future of work – *unless we do something*. The something should not be the palliative of fake jobs but rather the capacities, credentials and opportunities that have for too long been the prerogative of the middle classes. It is, for instance, a contemporary scandal that 45 per cent of our young people still leave school without a single 'O' level or its equivalent, branded as incompetent in a credential society before they have even started.[5]

BOX 1.4   HOW NEW IS THE DILEMMA?

In 1930 John Maynard Keynes, the economist, wrote:

We are being afflicted with a new disease of which some readers may not yet have heard the name, but of which they will hear a great deal in the years to come — namely technological unemployment. This means unemployment due to our discovery of means of economizing the use of labour outrunning the pace at which we find new uses for labour ... All this means in the long run that mankind is solving its economic problem .... I draw the conclusion that, assuming no important wars and no important increase in population, the economic problem may be solved, or be at least within sight of solution, within a hundred years. This means that the economic problem is not — if we look into the future — the permanent problem of the human race. ...

Will this be a benefit? If one believes at all in the real values of life, the prospect at least opens up the possibility of benefit. Yet I think with dread of the readjustment of the habits and instincts of the ordinary man, bred into him for countless generations, which he may be asked to discard within a few decades. ...

There is no country and no people, I think, who can look forward to the age of leisure and of abundance without a dread. It is a fearful problem for the ordinary person, with no special talents to occupy himself, especially if he no longer has roots in the soil or in custom or in the beloved conventions of a traditional society. ...

For many ages to come the old Adam will be so strong in us that everybody will need to do some work if he is to be contented. We shall do more things for ourselves ... only too glad to have small duties and tasks and routines. But beyond this we shall endeavour to spread the butter thin on the bread — to make what work there is still to be done as widely shared as possible. ...

When the accumulation of wealth is no longer of high social importance, there will be great changes in the code of morals. We shall be able to rid ourselves of many of the pseudo-moral principles which have hag-ridden us for two hundred years, by which we have exalted some of the most distasteful of human qualities into the position of the highest virtues. We shall be able to afford to dare to assess the money motive at its true value.

John Maynard Keynes
'Economic Possibilities for our Grandchildren'

'Give a man a fish,' the proverb goes, 'and you feed him for a day. Teach him to fish and you feed him for life.' It is enlightened self-interest for the privileged to invest in the under privileged, just as it is for the countries of the North to invest in those of the South. If they do not, they will each have to support the dependent population that they have helped to perpetuate and will have to live with the divided world that they have helped to create. The future of work in Britain is inextricably entangled with the saga of the classes, and no agenda should ignore that fact. This is no short-time project but a task for a generation. The planning and the preparation needs to start today if we are not to slide inexorably into a society that is more unequal, more unfair and more unpleasant than anyone would like to see. (Chapters 5 and 6 return to this theme in more detail.)

The one thing that can do most to frustrate that planning and preparation would be a refusal to accept that things are changing. Those who believe in the continuity of this industrial age and seek to cling to patterns of work and life as we knew them are not going to license or encourage any exploration of new possibilities. It needs courage to admit that the old must go to give place to the new, particularly if you have lived with the old all your life and know no other ways. Courage is fuelled by desperation or by vision. We must hope that we do not have to wait for the stage of desperation, despite a depressing tendency in the British to leave things that late.

The continuity of tradition is, for better or for worse, part of the British culture. It was summed up most appropriately at the General Synod of the Church of England a few years back, when the ordination of women, a momentous change for that assembly, was being discussed. A plaintive appeal was made from the floor in the middle of the debate: 'Mr Chairman, could not the status quo be the way forward in this matter?' But the British have another valued tradition, the ability to embrace the inevitable and to make it one's own. The establishment has long learned to pull any triumphant outsider into itself and, when necessary, to make his ways its ways. 'Co-option' it is called in sociological jargon, 'pragmatism' in politics. Typically this adaptive pragmatism seems to come into action when any marginal phenomenon rises above the 20 per cent barrier. Below that barrier any competition, movement or fashion can be discounted as a marginal nuisance, 'a little local difficulty' in Harold Macmillan's telling phrase. But once the 20 per cent credibility barrier is passed accommodation may be the better part of valour. When divorce affected only one in ten marriages, the divorcée was a criminal and a social pariah. As the 20 per cent barrier was approached, laws, values and language all changed. Divorce, if not always approved of, is now an accepted part of life.

Today 20 per cent of the workforce is part-time.[6] The formally self-employed are moving towards a 10 per cent share of the workforce. Unemployment is variously estimated at 12 to 18 per cent, depending on people's assumptions. On many fronts the traditional 20 per cent barrier in work is approaching. History suggests that there will be many attempts to push the figures back so that the groups once again become marginal and therefore unimportant, but that if they continue to break the 20 per cent barrier, they will be accommodated and adopted. That will mean changes for the other 80 per cent, and that's when society as a whole will start to change, sometimes fairly rapidly, as the inevitable becomes the accepted commonplace. The pain and the fury and the argument take place on the *approach* to the barrier, which is what we are witnessing today – the pain of the abandoned minority, the fury of their champions and the argument, of which this book is a part, as to whether this is a fundamental change or a mere blip on history's charts, a beacon or a flicker.

It is during this approach to the barrier that one is most conscious of the 80 and the 20 per cent. The 80 per cent are in favour of continuity. They hope and believe that things will right themselves and produce endless anecdotes and rhetoric to reassure themselves or to calumniate the others. To the 20 per cent the 80 per cent are 'selfish', 'uncaring', 'wilfully blind' or, more kindly, 'Micawbers' hoping that something will turn up, or just 'misguided'.

In these circumstances people turn, quite naturally, to reassurance from groups that think as they do, to newspapers and journals of their own persuasion, and cut themselves off from dissenting opinion. Society then becomes more divided, arguments more bitter and the 'group-think' phenomenon more common. In group-think situations groups of like-minded people move quite righteously, quickly and with total logic to extreme positions because dissenting views are excluded and each member reinforces the other. Some of the major political and military blunders of history (the Bay of Pigs is a frequently quoted example) are due to group-think decisions. Only by widening and legitimating the debate can the dangers of group-think and of a fractured society be avoided.

It is, of course, critically important to decide whether or not things are really changing. The continuity school, all deep within the 80 per cent, will look for temporary and piecemeal solutions to what they see as temporary problems. The language of job creation, supplementary benefit and one-year practical skills training all belong to the continuity school and make sense if everything reverts to normal in due course. If, however, we are contemplating the need to embrace a variety of 20 per cent groups within our continuing tradition of work, then these sorts of measure are mere palliatives, diverting attention and resources from the real problems. Those who sense a permanent change sometimes talk of structural changes in employment. They mean that the whole framework of employment may be changing; but in that it can sound like the replacement of one structure by another, the technical language gives false clues to ordinary people because it is the progressive destructuring of work that may well be the key change. If that destructuring does happen then we shall need to rethink much of the basis of the welfare state, which was anchored firmly to the employment organization and to full-time lifetime employment for everyone. That is why the future of work ends up as an intensely political agenda – and that, presumably, is why most politicians choose to stick tight to the continuity school. It is safer, and there are more voters in the 80 per cent.

The continuity school fears that by addressing this new agenda we invite the very problems we are seeking to solve. It would have us ignore these issues as long as possible in the hope that the world will revert to what it was or else that some kind of Darwinian process of evolution through survival of the fittest will sort it all out for us.

This has to be a dangerous and irresponsible view if any of the scenarios have any truth in them. Survival of the fittest condemns the least fit. Unemployment already divides the nation by accident of birth or training. Those who were born in the north or to poor or immigrant families, those who had the bad luck to join industries which were at their peak twenty years ago, those who find that they are 55 or 16 in 1984 – these could become, by the luck of the draw, the sacrificial victims to this kind of passive determinism.

If a society makes jobs the pivot of existence and then cannot provide enough jobs, or share out the available jobs more fairly, or find alternative pivots for life, it is practising deceit. Many people have written persuasively of the advantages of a society where employment does not fill the whole of life, where a few hours or a few people can earn enough for the many, leaving exciting opportunities for our discretionary time. That is a possible vision of the future, but we are at present asking the young, the unskilled, the handicapped and the immigrants to show us the way. It is unfair to them and unfair to the possibilities of the future.

2

The Job Scene

The reappraisal of work is forced upon us by what is happening to jobs. In this chapter we shall try to take a step back from the urgency and pain of today's lack of jobs and look at the changing pattern of work over the forty-year period from 1961 to 2001, for while the pace of the change can be and has been affected by government action, there are more inexorable forces at work. To focus only on unemployment as one aspect of this change and on the short-term past or future can be to blind oneself to the larger picture.

The bare facts for the United Kingdom are as follows (although the pattern in other European countries has not been dissimilar).[1]

(1) *There are fewer jobs than there were, but not that many fewer.* In 1961 there were 24.5 million people in employment in the UK, including the self-employed. In 1980 there were slightly more (25.2 million), but by mid-1982 the total had fallen to 23.5 million, where it seems to be sticking. In other words, jobs, contrary to popular opinion, did not start disappearing until 1981, and are still only 1 million fewer than in 1981.

(2) *But the jobs have continuously been squeezed out of business.* In the twenty years from 1961 to 1981 2.5 million jobs (13 per cent of all jobs) were squeezed out of the business sector (both private and public) in Britain, although output increased. The fall was constant but became most dramatic in 1980–1. Higher productivity has tended to mean fewer jobs.

(3) *Jobs were picked up by the public sector, however – but no longer.* From 1961 to 1980 all the jobs lost from the business sector were picked up by the state services sector, maintaining the total level of employment. This is no longer so. The state sector is now contracting (slightly) not expanding in terms of jobs.

(4) *In general, jobs have continuously moved from manufacturing to service.* In the twenty years from 1961 to 1981 3.3 million jobs

left the traditional worlds of mining, agriculture and manu-facturing, and 2.2 million jobs were added to the service sector (1.8 million of them in the public sector).

(5) *Meanwhile the labour force has been expanding. . .* The labour force grew by 2.2 million during the twenty years 1961–81. Therefore even if employment grew marginally up to 1980, *unemployment* would still have risen (and did rise). Statistically, this increase in the registered labour force was entirely due to an increase in the number of working wives (up by 2.7 million over the twenty years).

(6) *. . . and is going to go on expanding for the next twenty years.* The baby boom of the 1960s is now grown up and is beginning to work its way through the labour force, greatly outnumbering those who would normally be retiring. Over the next twenty years there will be an *extra* 1.5 million people of working age, most of them wanting jobs.

(7) *Demand for jobs has consistently exceeded the supply and looks like continuing to do so.* The *demand* for jobs has grown by 2.2 million people or nearly 10 per cent over the last twenty years and will grow by a possible 1.5 million over the next twenty; meanwhile the *supply* has contracted by 1.1 million (to mid-1981). The gap is unemployment – some of it temporary, too much of it permanent.

To put these trends in perspective, it helps to carry some pictures in the mind.

SOME JOB PICTURES

Picture no. 1: the three economies

The *market economy* is the marketplace of society, where things and services are made and delivered for a price. Most of the market economy is privately owned but not all of it. The railways, coal mines and steel mills of the nationalized industries are part of the market economy because they trade for cash. Some schools and hospitals, the private ones, are there, as are all the businesses, shops and firms which fill the high street.

The *state economy* is the collection of activities done by the state on our behalf, paid for indirectly by us through taxation. The armed forces, the police and Civil Service are part of this economy, as are the National Health Service the social services, local government, most teachers, doctors, nurses and the staff of most of our hospitals. It also includes all pensions and benefit payments.

These two economies make up the *formal economy* – the economy that is measured, forecast and used as the basis for estimates of our national wealth. All countries, whatever their political ideology, have a set of activities which are charged for at home and abroad (the market economy) and a set that are largely or wholly funded (the state economy). By and large the taxes from the market economy pay for the state economy (this is not wholly true because the employees in the state economy also pay tax), and by and large it is the market economy that earns the foreign currency which we need to buy food, raw materials and finished products from abroad.

The size of the state economy in relation to the market economy varies from country to country. There is no iron law which says what the relative size of the state economy should be. In effect, it can be as large as we are prepared to pay for in the form of taxes on the market economy. Britain's state economy is slightly smaller than that of her Continental neighbours.

The picture, however, includes a third economy, one that has been for long the unnoticed, taken-for-granted, younger sister of our society. It may well be a sign of the times that this third economy is now receiving the attention of economists, sociologists and political commentators. It will be analysed in more detail in chapter 3. Here a brief summary of this third economy will suffice.

The *informal economy* contains all the uncounted activity in which we engage. Just because an activity is not recorded or even charged for does not mean that it has no economic value. If I paint my own house, the labour content is the same as if I paid someone to do it for me. It is a payment to myself and therefore has no knock-on value. It does not become part of the currency of society, which is why government takes little interest in it. But, of course, if everyone painted his or her own house, there would be no jobs for house painters and less money circulating in society. It is this kind of possibility which is now stirring the imaginations of all sorts of people.

Strictly speaking, the informal economy has three parts:

(1) The black economy, the uncounted 'market' of the moon-lighter or the criminal. Although much talked about and much indulged in at the margin by all sorts of people, this part of the informal economy is probably not as large in Britain as many think – perhaps 3 per cent of the formal economy, much less than the estimated 20 per cent in Italy or 30 per cent in Eastern Europe.

(2) The voluntary economy, the voluntary work which occupies many of us for at least an hour a week (perhaps 18 million of us, according to one recent survey,[2] equivalent to 500,000 full-time workers).

(3) The household economy, the big part of the informal economy, containing all our domestic work, cooking, gardening, household maintenance, child care and old people care. It has always been there but unheeded and uncounted – perhaps because it has been largely woman's work? Recent estimates suggest that over half of the nation's productive work hours are spent in the household or around it. If this work were charged for, it might be equivalent to 40 per cent of the formal economy.[3]

In total, therefore, the value of the informal economy could be as much as half of the formal economy, although we shall never know precisely.

This first picture is important because it illustrates one aspect of the changing face of work in Britain: the contraction of the labour market and the expansion of the labour pool has meant that in 1980 3.2 million people were looking for work, with no room in the formal economy. No wonder that the informal economy suddenly becomes interesting and important. It is the only place left. Arguably, without the increase in the labour force the situation might have been containable. But we are not in the game of might-have-beens. The facts are clear. Jobs have continually been squeezed out of the market economy into the state economy and now into the informal economy where there is work of a sort, but no jobs and not much money. The question, to which we will return, is, will it continue like this, or can (should) the process be reversed?

BOX 2.1   SUMMARY DESCRIPTION OF THE HOUSEHOLD, FORMAL AND HIDDEN ECONOMIES

Type of personal gain	Whether part of hidden economy	Examples (some may fit several categories)
Working for no renumeration, often thereby avoiding paying others to carry out the work concerned.	No. This is outside the boundary of production. It is termed here the household economy.	Housework DIY Gardening Voluntary charity work Lift-sharing

BOX 2.1 continued

Type of personal gain	Whether part of hidden economy	Examples (some may fit several categories)
Working for money, fully declared as necessary to the Inland Revenue	No. This is termed here the formal economy	PAYE earnings, etc.
Enjoying a personal benefit and possible tax advantage from 'expense-account living', etc.	Yes	Enjoying five-star accommodation when conducting important business abroad on behalf of employer
Receiving remuneration of goods or services greater than their valuation for tax purposes	Yes – to the extent under-valued	Fringe benefits
Illegally submitting an incorrect or insufficient tax declaration	Yes	Undeclared income from second job. Self-employed undeclared gross income or over-stated expenses. Undeclared earnings (e.g. tips). Company tax evasion.
Frauds connected with the production of goods and services in the formal economy (not exclusively related to tax fraud)	Yes	Office pilfering. Fiddling of customers or employers by employee. Employment of employees 'off-the-books'. Shoplifting.
Undeclared criminal or immoral earnings	Yes, in principle	Drug trafficking Prostitution

Economic Trends, no. 313, February 1980

Picture no. 2: the four sectors

The second picture is of the *type* of work. In which sector of society does it occur? There are, traditionally, three main sectors:

(1)   agriculture and mining: the work of the early societies, pre-industrial societies, was mainly concerned with growing and extracting things from the land;

(2)   industry: the so-called industrial revolution saw jobs in agriculture diminishing and jobs in industry growing rapidly;

(3)   services: a category which includes all state services, health, education, and protection, as well as private service industries like retailing, banking and hairdressing.

To these many would now add a fourth:

(4)   information: to include all those activities which process or transform information, using pens, typewriters, computers, cameras, microphones, televisions and blackboards. It is really a sub-category of services but important enough in its potential to be separated out.

There is a traditional historical pattern to the changing importance of these four sectors, captured in the wave pattern shown in box 2.2 The facts are not in dispute:

(1)   the jobs in agriculture and mining in Britain went *down* from 1.7 million in 1961 to 0.9 million in 1981 (4 per cent of the total);

(2)   the jobs in manufacturing and construction went *down* from 10.2 million in 1976 to 7.6 million in 1981 (33½ per cent of the total);

(3)   the jobs in services went *up* from 11.4 million in 1961 to 14.7 million in 1981 (63 per cent of the total).

No official statistics are available as yet for the information sector in Britain, but Tom Stonier estimated it at 33 per cent of the total in 1971,[4] and Porat in the USA estimated that in 1967 the information sector of that country accounted for 46 per cent of the GNP,[5] although no attempt was made to relate that to jobs. Barry Jones, however, did attempt the breakdown for Australia, estimating that the jobs in the information sector in Australia in 1980 were about the

BOX 2.2   THE WAVES OF WORK

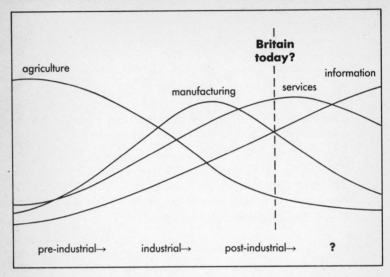

same as those in manufacturing. Here is the list he gives of the activities in the information sector:[6]

teaching	creative arts and architecture
research	design
office work	music
public service	data processing
communications	computer software
the media	selling tickets
films	accountancy
theatre	law
photography	psychiatry and psychology
post and telecommunications	social work
book publishing	management
printing	advertising
banking	the Church
insurance	science
real estate	trade unions
administration	parliaments
museums and libraries	

It is hard to avoid the impressionistic conclusion that many of these are the activities of the future, but at present they are concealed within the services sector or within the service bits of manufacturing. The reasons for the facts are not much in dispute either.

(1)   The substitution of capital equipment for labour in agriculture, mining and manufacturing has allowed output to grow while jobs decline (productivity, in these sectors, is a polite word for labour reduction).

(2)   The cheaper labour costs in the developing and newly industrialized countries have drawn traditional manufacturing industries away from Britain. (Coats Patons in 1981 made its own comparisons of the costs of its international labour force: UK 100, USA 117, Canada 133, Brazil 23, Philippines 10 and Indonesia 6.)

(3)   The long-term economic cycles of the sort described by Kondratiev (box 2.3) were probably exacerbated or accelerated by the jump in energy prices in the 1970s.

BOX 2.3   KONDRATIEV

Nikolai Kondratiev was a Russian economist who in 1926 pointed out that economic activity moved in 'long waves' of fifty years or so in duration. The first long wave was from 1789 to 1849 (going up until 1814, then down); the second from 1849 to 1896 (up until 1873) and the third from 1896 to 1945 (although Kondratiev only identified the upswing until 1920 and did not venture to predict the future). Carrying on his time sequence produces a fourth wave from 1945 to 1995; 1970–95 are the years of decline, leading into the beginning of a fifth wave at the end of the century.

Kondratiev did not think that the waves were accidental. 'Important discoveries and inventions' were made during the downswings and then applied in the upswings. In recessions Kondratiev's thinking receives renewed attention (e.g. by Schumpeter and Burns in the 1930s and by Kuznets in the 1970s), perhaps because he offers the promise of light at the end of the tunnel. Others, however, have had difficulty in fitting technological discoveries to his timescales, perhaps because it is extremely difficult to tie down precise dates and limits to technological invention. Did the computer start to be developed in 1834 with Babbage or with the first electronic computer in 1946 (the start of an upswing, incidentally)?

What Kondratiev did not comment on, but which is of perhaps the greatest social importance, is the fact that the principal focus of industrial activity tended to change with each successive wave.

The first wave, based on the steam engine, had its focus in England, which had at the end of it 40 per cent of the world's industrial production (compared with less than 3 per cent today). Activity then moved to Germany and northern Europe, with the development of chemicals, steel making and oil, and on to the United States in the third wave at the end of the nineteenth century, with the development of the automobile and the telephone. The fourth wave, after the war, focused on Japan and California and the emergence of electronics.

Meanwhile, as the waves of economic activity moved on, the societies and communities of the earlier waves had to adjust. Where will the fifth wave, if there is one, find its focus?

Britain is no longer primarily an industrial nation. Since the early 1970s, as Tom Stonier points out,[7] the gross profits of the service sector have exceeded those of the manufacturing sector. Our wealth now comes more from doing things than from making things. Should we be worried or not?

Box 2.4 makes it clear that this is not a peculiarly British situation. Jobs have been switching out of manufacturing all over the industrialized world for the last thirty years. In Britain and, to a smaller extent, in France and West Germany the flow was so great in the early 1980s that they could not all be picked up in the service sector, partly because of the restrictions on growth of the state economy and its services, partly, no doubt, because it was in these years that the anti-inflation policies of the Thatcher Government bit deepest into the industrial sector.

BOX 2.4   IS IT DIFFERENT IN OTHER COUNTRIES?

Answer: No, not really.

The last thirty years have seen a progressive fall in the percentage of the labour force employed in manufacturing in other countries, accompanied by enormous growth in the output of that part of their economies.

USA	1950	33%	1970	29%	1980	22%
France	1950	45%	1970	39%	1976	28%
West Germany	1950	48%	1968	48%	1976	35%
Japan	1960	28%	1970	28%	1976	26%
United Kingdom	1951	39%	1965	35%	1976	31%

Source: B. Jones, *Sleepers Wake* (Wheatsheaf Books, Brighton, 1982), p. 62.

BOX 2.4 continued

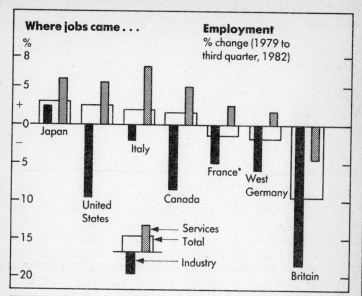

Where jobs came . . .

Employment
% change (1979 to third quarter, 1982)

% 8 5 + 0 − 5 10 15 20

Japan

Italy

France* West Germany

United States

Canada

Services
Total
Industry

Britain

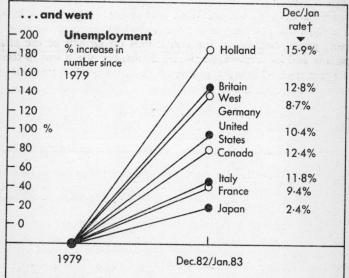

. . . and went

Unemployment
% increase in number since 1979

Dec/Jan
rate†

200		
180	Holland	15·9%
160		
140	Britain	12·8%
120	West Germany	8·7%
100 %	United States	10·4%
80	Canada	12·4%
60		
40	Italy	11·8%
	France	9·4%
20	Japan	2·4%
0		

1979   Dec.82/Jan.83

Notes: *Fourth quarter, 1981; †seasonally adjusted; ‡unadjusted.
Source: Economist, 18 September 1981 (OECD figures).

But if the past and the present are not in too much dispute, the future is. No one can be too sure which way it will go and what policies will help or hinder it. If you were to turn back to box 2.2 and block out the chart after the dotted line representing Britain today, would your estimate of the future fit the waves on the chart?

Picture No. 3: the three fingers

The job situation can be most conveniently summed up by box 2.5, in which the situations described in pictures 1 and 2 are summarized in just three lines. In broad terms the lines are the same for all industrialized countries, although the detailed numbers change from country to country.

The top line represents all those people of working age. Not all of them will choose to apply for jobs. A few will not be able to because they are too ill or in prison. This is the *potential* workforce. It is growing and will go on growing until the end of the century and beyond. The birth rate may have dipped in the 1970s, but there will still be more people entering the workforce than leaving it. We are en route to a middle-aged society, comprising people of working age, more and more of them wanting work.

The middle line represents total employment. By and large this has held steady apart from a sudden dip in the early 1980s in Mrs Thatcher's Britain.

The bottom line represents the jobs in agriculture, mining, industry and construction, the 'old economy'. These jobs, in all countries, are becoming fewer, even if output holds up. Professor Tom Stonier estimates that this line could drop to 10 per cent of the total early in the next century.[8] Even if it does not drop that low that fast, it is still declining everywhere.

The space between the middle and bottom lines includes all the jobs that are not in agriculture, mining and manufacturing – the service sector, both public and private.

The job scene, then, is one of three fingers pointing into the future. The top line of potential workers points slightly upwards; the bottom line of traditional occupations points downwards. So far, with occasional lapses, the middle line has held up. The result has been not the collapse of employment but a growing gap between that middle line and the top line. Even under that middle line there is a lot of potential space. Of these jobs 20 per cent are part-time, and 10 per cent are self-employed, which can mean very full-time work but also much less than full-time. A survey by PA Management Consultants of thirty British companies found that if they cut working time by 10 per cent, they would expect to need to increase their staff by only 1 per cent. In other words there is still a lot of slack in full employment. To put it all another way, if we wanted to provide full-time jobs for all the workforce, we would need to create 8 million more jobs, as many jobs as there were in all of agriculture, mining and manufacturing in 1983.

BOX 2.5 THE JOB FINGERS

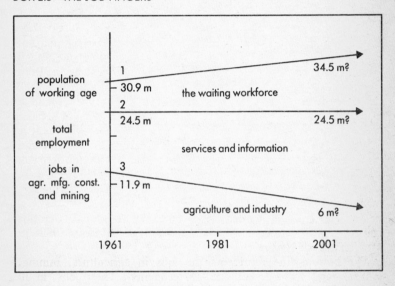

THE QUESTIONS THAT MUST BE ASKED

These pictures of the changing job scene inevitably raise some big questions for the ordinary citizen.

Is the decline in manufacturing jobs inevitable?
Will it continue?

It is hard to see why it should not. That does not mean that the *output* of the manufacturing sector should decline, any more than agriculture's did, although it may consist of different products. There is, after all, no shortage of things that people need and will need. Nor do we need to manufacture only for export. There is enormous scope for important substitution, making in this country the things we commonly import, but if we are to make and sell more, we first have to reduce the cost. This is the iron law of international comparison.

International comparison means that, in a free trading world, you can continue to sell your increased output only if it is produced at a lower cost, which in the long run means with less labour. New industries and new products can escape from this inexorable pressure of efficiency for a time until competition catches up, but not for long. Monopolies, of course, are exceptions and are therefore unpopular. The iron law of international comparison requires that you improve your efficiency, cut your costs, by something like 3 per cent every year *just to stand still*. Therefore to keep your staff at the same numbers you would have to grow by 3 per cent p.a. in terms of output. To keep your staff and pay them more in real wages you would have to grow faster than that – by at least 5 per cent p.a. Individual firms can do this for a while, but for the whole of manufacturing industry it is an unlikely prospect. Even Japan managed it only for a number of years and is now cutting her manufacturing workforce along with the rest of us.

Government economic policy can in the short term produce kinks in this decline. International competitiveness can be improved or hindered by the state of the currency; domestic demand can be boosted or dampened down; wages are affected by inflation and by bargaining arrangements. Unfortunately, it is difficult, if not impossible, to push all these variables in the right direction at the same time. A cheap currency helps exports but fuels inflation. Reduced wages keep labour costs down but diminish demand because people have less money to spend. Economic policy has to be a continually adjusting compromise. In the long term the shift does seem to be inevitable. Capital drives out jobs, or at least moves them somewhere else. We could have passed a law outlawing tractors or requiring everyone to be accompanied by a man on a horse. We wisely did not. We could pass laws outlawing robots or require them to be surrounded by men. If we are wise, we will not.

There are many, however, who do not accept that the shift of jobs out of manufacturing is inevitable. Some put their faith in Kondratiev (see box 2.3) and look for the next up-turn, believing that new technologies will trigger new businesses and new products yet unheard of. Others would go the other way, wanting us to find labour-intensive processes and intermediate technologies. Some look to a revival of Britain's industrial and competitive skills and energies to give her back some of her market shares of the nineteenth century. Others believe that if we all worked shorter hours, we could keep enough jobs.

What should we make of these suggestions? New products and new technologies are crucial, but where is the evidence that they will produce more rather than fewer jobs than the businesses they replace? Kondratiev was talking about *output*, not *jobs*. As long as manufacturing was labour-intensive the two were the same. Now they are not. The new businesses think in terms of hundreds of people, not thousands, even when they have big new plants. Similarly, any British manufacturing revival will have to be based on high added value, high mark-up products if they are to be successful because we are, by comparison with others, a high-wage economy. We must, therefore, become a low labour economy – fewer jobs but well paid ones. And that, too, has to be the ultimate answer to those who look for more labour-intensive intermediate technologies: we will not be able to sell the goods they produce because they will cost too much if we pay the labour properly.

The industries of tomorrow will include biotechnology (replacing parts of the chemical and pharmaceutical industries), micro-technologies (which will allow you to programme your kitchen, help the blind to see and keep petrol consumption in cars to a minimum), fibre optics and communication satellites. Will any of them be using a cheap or extensive labour *in the manufacturing process*? They may need more people to *use* them, but not to *make* them. Finally, those who would cut down the hours so that the same number of people can have the same number of smaller jobs do not always realize that they must also reduce the wages. A four-day week which costs the same as a five-day week makes no difference to the wage bill and does not improve efficiency.

Where will all the jobs go?

As agricultural labour was phased out, the jobs shifted to industry, although it may have taken us a little time to wake up to what was happening. As Barry Jones notes,[9] it was 1884 before the term 'Industrial Revolution' was first used in English by Arnold Toynbee. As *industrial* labour is phased out, where will the jobs go to next?

The services is the conventional answer provided by many, and indeed, as we have seen, the service sector has by and large picked up the pieces, the major increases being in state education, state health, private banking and private insurance (the only two industries to have produced more jobs in the last ten years).

Information, as we have also seen, is another possible answer, identifying the information-providing part of the service sector as the growth element (see p. 22).

The personal services is a third possible answer – the services involved in providing people with food and drink, entertainment, travel and holidays, health, education, sport, heat, water, cleaning and maintenance (see p. 7). There is obviously some overlap between personal services and information services. Education, for instance, is central to both; it is inextricably involved in information in all its guises and at the same time is one of the most important of the personal services. But, in spite of the overlap, there are differences between information and personal services which make it worth living with the confusion between the two for a while.

Are information services the path to the future? To some, like Professor Stonier, they are the only path, particularly the education services. To others, such as Clive Jenkins and Barrie Sherman,[10] they are ripe for automation. The labour element in many of the service industries is just that – hands and eyes shifting information. It can, with relative ease, be replaced by 'clever machines' in banks, check-out counters, libraries, insurance offices, stockbrokers, even eventually in doctors' surgeries and the classroom. Jonathan Gershuny foresees the possibility that no one who is not a professional will be employed in the services.[11]

Arguably, the information services are as prone as manufacturing to the replacement of labour by capital. On the other hand, it is in this area that it is easiest to see the proliferation of new products and new businesses.

On balance there seems a good chance that the growth of new products could be enough to compensate for the loss of jobs in the older services. Any major expansion, however, would have to involve, as Professor Stonier argues,[12] considerable growth in the provision of education, and this would almost certainly need to be state-financed if it were to be available to all, thereby continuing the shift from private to public sector.

Some of the personal services are also susceptible to decimation by automation, but less so. Water, gas and electricity have already gone down this path; transport would like to go further than it has gone, with self-drive trains, for example, as would retailing, with automatic check-outs. The police are still labour-intensive, but the armed forces are less so. Cooking, cleaning and maintenance of all sorts are in the middle of a small-scale revival now that they are done by small, independent businesses rather than by servants; yet here, too, the capital equipment of a cookery partnership or a garden contractor is considerable and dispenses with the need for many, if any, employees.

Nevertheless, as long as we have waiters there will be a limit to the numbers they can wait upon; hairdressers can cut only so many heads of hair, directors can direct only so many actors and so many productions. Many of the personal services *do* increase their jobs as fast as they increase their output because they are labour-absorbing not labour-saving. A play without actors is not a play, and a hospital without nurses would be a strange place, no matter how stuffed with equipment it was.

There is also, in a strange way, scope for substituting labour for capital by making part of manufacturing into a personal service. If that point seems obscure, imagine what would happen if expensive, long-life cars became fashionable again, so that Bugattis and Lagondas roamed the roads once more in modern guise. Their manufacture would be highly automated, but their long life would be ensured by continuing maintenance (personal service) in multitudes of garages. If the 'replacement society' became the 'repair society', there would be an increase in high value-added manufacture (which would suit the British) and an increase in personal service maintenance.

BOX 2.6   THE CHANGING BRITISH AND AMERICAN JOB SCENES

Britain		
How jobs will rise and fall	*1980* (000s)	*1980–90* (+ or − %)
Managers, administrators	2,129	+ 5.7
Education professions	984	− 4.3
Health professions	986	+ 9.4
Other professions	562	+11.5
Literary, artistic, sport	447	+26
Engineers, scientists	576	+14.3
Technicians, draughtsmen	601	+12.5
Clerical	4,056	+ 0.5
Sales	1,417	− 5.6
Supervisors, foremen	104	−10
Engineering craftsmen	2,143	− 5.3
Other transferable craftsmen	907	−18
Non-transferable craftsmen	675	−27.4
Skilled operatives	622	−15.3
Other operatives	4,712	−18
Security occupations	386	+25
Personal service occupations	2,932	− 3
Other occupations	789	−38.1
Non-manual	11,755	+ 3.8
Manual	13,271	−12.9
All occupations excluding HM Forces	25,026	− 5

Source: University of Warwick Institute for Employment Research, using Warwick occupational categories, July 1983.

BOX 2.6 continued

The USA

Prospects	1980 (000s)	1990 (000s)	1980–90 (+ or −)
Ten best prospects			
Secretaries	2,469	3,169	+700
Nurses' aides	1,175	1,682	+507
Janitors	2,751	3,253	+502
Sales clerks	2,880	3,359	+479
Cashiers	1,993	2,445	+452
Nurses	1,104	1,542	+438
Truck drivers	1,696	2,111	+415
Fast-foodworkers	806	1,206	+400
Clerks	2,395	2,772	+377
Waiters	1,711	2,072	+361
Ten worst prospects			
Postal clerks	316	310	− 6
Clergy	296	287	− 9
Shoe machine operators	65	54	− 11
Compositors and typesetters	128	115	− 13
Graduate asisstants	132	108	− 24
Servants	478	449	− 29
College teachers	457	402	− 55
High school teachers	1,237	1,064	−173
Farm labourers	1,175	940	−235
Farm operators	1,447	1,201	−246

Source: US Department of Labor statistics, *Time* magazine, 30 May 1983.

We must, however, be realistic about one thing. Service jobs in small businesses will not be paid as well as jobs in capital-intensive industry. The labour market will work effectively in these new areas because the jobs will be less well organized and less clearly defined. An effective labour market may produce more jobs, but it will also keep wages down. The personal service sector is likely to be a bit of a jungle.

Will there then be enough new jobs in information or personal services to pick up the continuing fall-out from manufacturing? No one knows. That will depend on a number of considerations.

(1) The wealth of the nation. Many personal services can be provided by ourselves for ourselves in the household economy. You have to feel rich to employ a gardener, hire a video, use a solicitor or travel on holiday. Where does this wealth come from? It has to come from an efficient, and therefore lean, business sector, both industrial and commercial.

(2) The attitude of the Government. Any major expansion in education or health (the traditional growth areas for jobs) has to be in the state economy. Private activity, although important, will never be enough to challenge the state as the main provider of jobs in these areas, even if we include under private activity health and beauty salons, gymnasia and diet books as well as private beds and private nursing, or home computers, video-recorders and do-it-yourself books as well as private schools.

(3) The growth of an entrepreneurial spirit in the services. It is easier to start a new enterprise in the services than in manufacturing. You need less plant and equipment, less capital and fewer customers. You can start by employing only yourself at the end of your telephone. But it still requires a cast of mind that thinks in terms of a 'business opportunity' rather than a 'job'.

(4) The understanding that the level of wages in small businesses does not at first reach the level of wages in the capital-intensive manufacturing sector.

Even then there is no guarantee that the increase in services will be able to deal with the outflow of jobs from industry. Even if it does, even if that middle 'finger' in box 2.5 representing total employment keeps to a level of around 24 million until the end of the century, the number of disappointed job seekers will grow because the top line, representing the available workforce, will continue to grow, by 1.5 million people between 1981 and 2001.

Three things seem certain, therefore: the new jobs will come from information services and particularly from personal services; there will not be enough of them to return us to full conventional employment; they are unlikely to be as well paid, for a time at least, and may even be exploiters of casual labour.

Is there nothing that central government can do?

Yes, there is much that the government can do. Some of it is already happening.

(1) The government could be a bigger *investor*. The infrastructure of the country could absorb large sums of money. The railways, roads, sewers and telephone cables of the country badly need modernization. Too many of our houses are, in effect, museums of the Victorian era. Hospitals need equipme buildings. Schools in parts of the country have seen no maintenance for a decade or more. By spending mone things the government would not only repair our in

but would also provide jobs for private industry and private individuals. Because such spending is capital expenditure and, inevitably, spread over time, it would not fuel inflation or pump up wages or prices, but it would create jobs.

(2)   The government could be a bigger *employer*. The health service could do with more doctors and nurses. The schools and colleges may be enrolling fewer pupils, but that need not mean fewer teachers if the money were available. Even if, in the interest of efficiency, the administrative component of the state economy – the clerks, secretaries and statisticians – should not be increased, there are good arguments for increasing the numbers of people in the front line so that more people could get more help more quickly and more directly in all spheres of life. The supervision of our recreational facilities, the aftercare of offenders, the management of our woodlands all need relatively unskilled workers and could be done better with more of them.

(3)   The government could be a better *sponsor*. The country could do with more better-educated people, especially in the new technologies. More research and development are required to create the products and the ideas which will be the foundation of the new businesses we need. Some of those new businesses will find it hard to collect enough capital to lift them to a viable size. In all these areas the government could help – by providing grants to more students to study both here and overseas, by sponsoring more of the long-term pioneering research that individual firms find hard to justify and by financing more seedling firms.

But all of these initiatives would cost money and would not create quite as many new jobs as people sometimes think. The sums are not in too much dispute. Something like £10,000 per year will create one job, whether that £10,000 is spent in the form of capital, as payment for work on the infrastructure, or as salary to a new employee in an office, hospital or school: £10,000 million extra each year would create 1 million jobs. That is a lot of money, but before we decide that it is worth spending it, let us consider what sort of jobs they might be. Some might be professional jobs in teaching, medicine or research. But how many teachers could, physically, be absorbed by the system? If we add 10 per cent to existing numbers, we get only 44,000, and it could take three years or more to train and place them. Even if the state economy *as a whole* employed 10 per cent more people, that would add only 520,000 to the payroll. Could the construction and electrical engineering industries absorb the remaining 480,000 people even if the money was available? It would mean something like a 25 per cent increase in their manpower – numbers that they would not need to do all the work that they might conceivably be asked to do. To insulate all the houses in Britain, the Select Committee of the House of Lords on Unemployment calculated,[13] might occupy 3,000 people for ten years. A big civil works programme on railways, roads and sewers might need, the Committee thought, 62,000 people p.a. A far cry from 480,000.

Let us, however, be wildly speculative and say that a free-spending Government might, both directly and indirectly through economic spin-off, create those extra 1 million jobs gradually over the next five years (because it could not happen immediately), and let us suppose that the population were willing to pay the extra 8p in income tax that such a programme might require or to accept a doubling of National Insurance contributions, *even then* we should not solve the problem. By the end of the decade there would still be 3 million workers unemployed, as a minimum, because of the extra numbers joining the workforce.

This is not an argument against government action on jobs. As we shall suggest in later chapters, government needs to do some of this and much else besides. It would, however, be dangerously naive to think that the Government, any Government, could buy enough jobs to solve unemployment.

What would happen if we were left with no industries?

In principle it would not matter if Britain manufactured nothing, and all goods were imported from other countries. But we would then have to export enough of something else, our services or our North Sea oil, in order to pay for them. In practice it is unlikely that we could ever sell enough of our financial services, our design or engineering consultancies, our education or our tourist potential to pay for all we need. We certainly do not do that now. North Sea oil has bailed us out – but for how long? We need a very efficient industrial base (i.e. employing few but precious people) both to earn our foreign exchange and to make the things, be they cars, washing machines or word processors, which otherwise we should have to import. As a leading US industrialist has remarked, 'A country cannot live by selling each other hamburgers.'

Countries, however, do not disappear if they cannot pay their bills. They only become poorer. In the worst scenario the situation is self-correcting in a rather depressing way. If we had nothing to sell abroad, we could buy nothing abroad. The pound would fall in value so far that we could not afford to import anything. Everything would have to be home-made with home-made materials. Industries would start up again because we would all *have* to buy British goods no matter how expensive or undesirable they were. Jobs, of a sort, would spring up again. Life, of a sort, would go on. But it would be a life devoid of choice, a life without much discretionary income; we could not travel, drink French wine, own a Japanese television or almost any kind of camera. It would be a dull and depressing world, but there would be jobs.

To avoid this dreary scenario we need wealth – wealth to spread around, wealth to maintain the relative affluence of the post-war decades and wealth to spend abroad on imports and raw materials. Wealth, however, does not come only from manufacturing – that is a conceit of the industrial age, just as it was a conceit of the agricultural age that wealth meant land. Essentially, you create wealth by giving value to something that formerly had no value. Tilling the land and creating crops where there were once weeds; turning the lathe and producing finely machined parts from crude steel – these are traditional wealth-creating activities, but today we can also apply ingenuity and expensive technology to the floor of the North Sea and release its oil, dormant and valueless for centuries. We can use the human brain and a computer to produce a program where nothing existed before, combine chemicals to make a drug to cure obesity, depression or arthritis. We could devise new forms of food or energy, new ways of educating people, new systems for passing information, new procedures for insurance or investment. The 'things' which we sell do not have to be tangible, or even visible. Nor do we necessarily have to use a lot of people to make them. The value of something depends on what people will pay for it, not the cost which goes into it. It is another fallacy of the industrial age that price is related to cost – that the something which costs little to make can be sold for little. Where, however, the basic ingredient is not physical effort but knowledge or talent the equation breaks down because we do not know how to cost knowledge. A painter charges for his talent, not his labour. Kuwait has the richest citizens of any country, not because they are hard-working or particularly talented but because they are lucky; they have a resource which other people need and whose price is unrelated to its cost.

The alluring future for Britain is based on this 'painter's wealth' – money paid for talent, not for time, a very sophisticated industrial base producing prosperity but very few jobs and all those jobs at the professional or technical level. An alternative future sees Britain as an industrial landlord, allowing herself to be used by foreign manufacturers as their European base, producing relatively conventional products very efficiently, using robots rather than people. The depressing future sees Britain isolated from the world, struggling to be self-sufficient and rebuilding her industrial base. In other words, we need an industrial base, but we need it to produce wealth, not jobs.

THE SHORT ANSWER

The short answer to the question 'Will there be full employment again?' has to be 'Not in our lifetime.' Manufacturing, like agriculture, must become a source of wealth but not of jobs.

The services could pick up much of the lost employment if the country is rich enough to afford a growing state service sector and more of the proliferating personal services which follow from education and affluence. The informal economy will pick up the residue of jobs and turn them into uncounted work. The poorer we are, the more we will do for ourselves; the richer we are, the more we will be tempted to use the personal services, but the work in the informal economy and in the personal services is likely to be insecure and poorly paid for most. That could be a very depressing conclusion. Jobs and affluence for the few, self-sufficiency and getting by for the many.

That, however, is to think too conventionally. It is to believe that a 'job' or 'employment' is the only way of working and the only way to distribute money and status in society. If jobs provide the skeleton for our lives, then, of course, we fall to pieces when the skeleton is removed. Should society be so job-fixated? The next chapter will argue that work is already being redefined, that the job is no longer quite what it was and that there is an array of possibilities if we start to play around with the conventional notion of the 100,000-hour lifetime job.

3

Rethinking Work

'Everyone has the right to work,' says the Universal Declaration of Human Rights, signed at the United Nations after the last war. At the time that meant the right to a job, but, as chapter 2 argued, no country can deliver on that promise any more. Can 'work' therefore mean more than a 'job', or must we start rationing work as if it were some scarce commodity? Rationing work has to be crazy, given all that needs to be done in the world and given people's seemingly innate need to express themselves through activity of some sort. Why ration something which should be limitless? That is the 'lump of labour' fallacy – the idea that there is some fixed finite amount of work to be done in the world, which has to be rationed out to those who want it.

Because the rationing of work is crazy, we are already beginning to rethink the whole concept of work: we are rethinking what we mean by work; why we want or need to work and the way in which work is organized. Is there a work ethic any more? Do members of the younger generation see work as central to their lives? Recently 38 per cent of West Germans replied to a survey question that they would be glad if 'less emphasis were placed on work in the future'.[1] Are other countries any different? Should we despair or be relieved?

What do we mean by work? I once asked a young woman what she did. 'I write television plays,' she said. 'Unfortunately, they never get accepted.' 'What do you do for money then?' I asked. 'I pack eggs on Sundays,' she replied. Which was her work – her job or her interest?

To the ancient Greeks work was demeaning, but by 'work' they meant physical toil done at the behest of someone else. Study, politics, military service, even digging one's own garden were not regarded as 'work' because they were done from choice. Are we getting more like them, using 'work' to mean the things we have to do and 'interest' or 'leisure activity' to signify the things we like doing? Is work going out of fashion?

The ancient Greeks had slaves to do their 'work'. Will we have robots, or is that science fiction? If we do not have jobs to go to, necessary work to do, how will we occupy ourselves? The free Greeks were a small elite, enough to fill a small market town, not a nation of 35 million adults. How will we get the money to live on if we do not have jobs? The free Greeks lived on the interest from their wealth. Is there any way in which 35 million people can do that?

Jobs themselves are changing. Of the workforce 30 per cent now are part-time employees or self-employed, perhaps more from necessity than from choice, but we may come to rejoice in the inevitable, for what evidence there is suggests that the part-timers and the self-employed are the most content with their work.[2] Will the process go further? The Government already provides incentives for the unemployed to become self-employed. Why not classify all unemployed as self-employed with subsidies for those who need them? After all, the word 'unemployment' entered the language only in the nineteenth century. Perhaps it could disappear in the twentieth if not having a job no longer meant not having work.

An insurance company in France allows its staff to work at home one day a week. Rank Xerox has turned fifty of its key head-office staff into 'networkers', independent consultants working from home. Is this another portent of the future? Will big organizations look more like networks then pyramids? Perhaps the products of the future will come with a list of people's names attached, like the credits that follow a television production, everyone with a mention.

What is automation going to do to us? One microprocessor can replace more than 600 separate parts in an assembly, and the microprocessor can be made and fitted by a robot. A word processor that does the work of three typists already pays for itself in eighteen months and gets cheaper every month. They need brains to design them and fingers to work them but nothing much in between. What does that mean for the skilled worker, for the well trained apprentice and the men of muscle, with jobs that are either too big for them or too small? Or will those new clever machines be tools for all of us if we have but wit enough to see them as such?

It will be the argument of this chapter that the seeds of the future lie in the present. It is at the edges of things, among the 20 per cent, that we can glimpse what lies ahead. Society is already beginning to adapt to the changing shape of work. It adapts grudgingly, slowly and ungraciously because the changes are being forced on it by economics, by technology, by the harsh realities of business and the need to survive. Surprisingly many people find that the water is not as cold once they are in it and that the new concepts of work would have much to commend them, particularly if they were accepted more openly by society and if society's laws and regulations took more account of them. This chapter therefore will look at the new things that are happening. They are not happening everywhere. They are not happening to everyone. They are not the complete recipe for a new society. They still affect the 20 per cent, not the 80 per cent, but they are pointers to the way in which things are going, and they do have implications for how all of us, the 80 per cent as well as the 20 per cent, may have to adjust our thinking about work. We shall look in turn at:

new worlds of work;
new meanings of work;
new patterns of work;
two options

It will be clear, first of all, that there is a lot more work going on in society than is officially counted and recognized. What would happen if we made these other types of work more respectable, more noticeable and more legitimate? It might help a lot to ease the social

BOX 3.1   THE MICRO'S AXE

The Type I automatic system for switching telephone calls employed twenty-six people to produce a unit of a certain capacity. It is estimated that a fully electronic System X unit of the same capacity could be made by only one person.

Siemens has trebled its output of electronic telexes but cut its employment by 20 per cent.

In Sweden's Saab-Scania car plant a robot that replaces three workers pays for itself in eighteen months.

The 'Office 1990' report by Siemens suggests that over 40 per cent of office tasks will become standardized and 25–30 per cent automated out of existence.

FIET, the umbrella organization for the clerical trade unions, has estimated that 5 million of Western Europe's 18 million office workers could be redundant by the end of the year [1981].

The Simon Nora–Alain Minc report to the French Government foresaw a 30 per cent job loss in insurance [during the course of the 1980s] and a similar saving through 'wastage' in banking.

Standard Electronic Lorenz switched to electronic telex machine production, replacing 936 parts with one microprocessor.

Giles Merritt
World Out of Work (Collins, London, 1982)

and psychological stigma of unemployment if its effect were perceived as releasing people for the other sorts of work, but that redefinition would conveniently slide over the problem of money. No doubt many people would be delighted to spend more time at home, would love to try to make a small business out of their spare-time interests or work in the community for free *if they could afford it*. Redefining work can too easily be a charter for the pensionable executive or civil servant with savings and a mortgage-free home but for few else, which is why any redefinition of work has to be linked to a redefinition of the job and to the practicalities of money.

NEW WORLDS OF WORK

Jobs belong to the formal, official economy, but, as we have seen already, there is also the work of the informal, unofficial economy. 'The death of the official economy has been much exaggerated,' says Professor Richard Rose of Strathclyde University,[3] going on to point out that working in the official economy is still the norm in every Western nation today. It is, in fact, impossible to live outside this official, formal or 'white' economy in the modern industrial state. No family can today avoid going to a supermarket for some of its food. No home can do without some capital equipment bought in the formal economy. All wage earners – and they are still the great majority – are in the formal economy. Nevertheless it is no longer true that the only world of work is that of the formal economy. Twenty years ago work, to everyone, meant working in an organization; employment implied a job. Fewer than 7 per cent of Britons were self-employed, a smaller proportion than in any other nation, and they were mostly in the professions or were running small shops, still pillars of the formal economy. It is different today. The economy has many colours and many shades. We have to understand the full spectrum of these colours if we are to understand work, because so much of work goes on in the different parts of the informal economy.

The informal economy is hardly invisible, but it has attracted more anecdote than measurement. Nonetheless, as we saw in chapter 2, it would probably be equivalent to half of the formal economy if we were able to put cash values on all the activity that goes on in that uncounted economy. It remains uncounted because it is of little obvious value to government. Where no cash passes or is acknowledged to pass, no taxes can be collected, no national insurance charges levied. To government the informal economy is a drain, not an asset. It is even a hazard; for where economic activity is not registered, be it in the garden shed or in the home, safety measures cannot be enforced, nor the rights of workers protected. This informal economy is growing in importance in spite of government, partly because in a recession it is cheaper to do what you can for yourself but mainly because work, like water, finds a way through in the end, and the informal economy is a wide-open market place for personal skills and services, whether they be sold or given for free.

The informal economy is important to our analysis of the future of work because it is the residual economy; it is the place where society's unused demand for work ends up. It is the reservoir of work. Reservoirs may look like pools of idle water, but they are recognized to be the right way to channel and conserve an excess of supply. Society would be wise to see the informal economy as a reservoir rather than a sump. This reservoir naturally fills up in time of slack demand and an increased supply of people. More and more people are, therefore, conscious of the different worlds of work which it comprises, for the full economy today has many shades:

(1)   the black economy – the illicit (because undeclared) market activities of small businesses and self-employed people;

(2)   the mauve economy – the burgeoning personal services and home businesses on the fringe of the formal economy;

(3)   the grey economy – the completely legal but uncounted domestic and voluntary work in which everybody is engaged to some exent.

We need to look at each in turn because they are very different: it confuses many issues to lump them all together.

The black economy

This is the best-known of the three informal economies but not by any means the largest. How large is it? No one knows for sure, of course. Looking at the estimates of expenditure and comparing them with estimates of reported income is one way. This suggests that people are spending 2 per cent more than they admit to receiving. Dilmot and Morris would support a figure of between 2 and 3 per cent.[4] On the other hand, the Inland Revenue's informed guess of 1980 put it at around 7½ per cent. This would imply that the average household spends about £1,000 a year in the black economy. That feels high, but we shall never know precisely. What is clear is that undeclared cash transactions are a familiar part of our national life. They are not regarded by the customer as wrong – as in the case of prostitution, it is he who lives off immoral earnings that is to blame, not he who pays. As long as this attitude continues the black economy will be a persistent and pervasive part of our economy, although not perhaps as large a part of it as the media would have us believe.

We are not alone. The black economy features in every other country's statistics except for those of France, where officialdom insists that it is negligible. In some places (Italy and part of Eastern Europe) it is a significant part of the economy, keeping some individuals in comfort but impoverishing the government by denying it tax revenue on a large lump of economic activity.

BOX 3.2   THE BLACK ECONOMIES

The USA
Professor Peter M. Gutman of the City University of New York estimates the underground economy to be 10 per cent of GNP. The US Internal Revenue Service put it between 5.9 and 7.9 per cent of GNP. It calculates that 4.5 million people live entirely on earnings from unofficial jobs.

Italy
There are thought to be 3 million to 5 million Italians engaged in *lavoro nero*, double the 1.8 million unemployed. They are thought to produce the equivalent of 20 per cent of GNP.

West Germany
Schwarzarbeit is thought to have grown fivefold in as many years, with 2 million workers accounting for a hidden 2 per cent of GNP. According to *The Times* of 3 March 1981, 70 per cent of carcass work is done by illegal butchers and 90 per cent of painting by illegal painters.

France
A fairly small black economy occupies only about ½ million people according to the French Information Centre, although *Tthe Times* puts the figure higher.

OECD
The International Labour Organization estimates that between 3 and 5 per cent of workers are unregistered.

The USSR
Different observers put the size of the black economy at between 20 and 25 per cent.

Gerald Mars
Cheats at Work (Allen & Unwin, London, 1982)

Who works in this black economy? Not everyone has the opportunity. Professor Richard Rose estimates that only 11 per cent have good opportunities, while another 40 per cent have some opportunity,[5] but that leaves half of the workforce with no opportunity of working in this unofficial marketplace. After all, as Rose points out, the goods and services supplied must be ones that do not require complex accounting systems, bureaucracies or technologies, as do banking, insurance and the telephone service (you cannot do a bit of banking on the side on your own). They must

BOX 3.3   WHO COULD WORK IN THE BLACK ECONOMY?

Employment according to likely opportunities to work in the unobserved economy

	Number (000s)	% total
Relatively high		
Construction	1,265	5.6
Textiles, leather and clothing	813	3.6
Agricultural, forestry, fishing	370	1.6
Total	2,448	10.9
Relatively low		
Professional and scientific services	3,717	16.5
Distributive trades	2,790	12.4
Miscellaneous services	2,519	11.2
Total	9,026	40.1
Virtually nil		
Engineering and allied industries	3,121	13.9
Public administration	1,596	7.1
Transport and communication	1,500	6.7
Rest of manufacturing	1,322	5.9
Insurance, banking and finance	1,258	5.6
Food, drink and tobacco manufacturing	681	3.0
Chemicals, coal and petroleum	470	2.1
Metal manufacturing	401	1.8
Mining and quarrying	344	1.5
Gas, electricity and water	347	1.5
Total	11,040	49.4

Source: Compiled by Richard Rose from Central Statistical Office, *Social Trends*, no. 13 (HMSO, London, 1982), table 4.7; data for 1980.

be products that can be produced 'invisibly' rather than in factories that can be seen by tax inspectors, and they must require only the kind of capital equipment that can be carried away or stowed in a corner. Moonlighting cannot be everyone's work. Rose, in fact, compiled a table (box 3.3) indicating the opportunities that are likely to be available.

There is a curious sense of morality among many moonlighters who do not see the need to declare their extra earnings ('This is my money, my personal pocket-money, not part of my regular income'). It is seen by many as a legitimate extra, outside the taxman's scope. Legitimate extras are one thing, but when people operate full-time in the black economy then society is being substantially cheated. There are holes in the *autostrada* in Italy because the state cannot collect sufficient taxes from the formal economy to pay for essential public works, while many an Italian lives tax-free on his black economy business.

As more sophisticated tools and materials become available, it will be more and more tempting for individuals to 'do a little on the side', for more and more of those in the potential black labour market to augment their earnings. This world of work seems bound to grow. Should it be stamped out (but how?), or should moonlighting be seen as the seed of entrepreneurial activity, as a legitimate extra, as personal earnings untaxed until they reached a certain level? Perhaps the state has nothing to lose by making the inevitable respectable, on the grounds that it is not going to collect the taxes anyway, that if moonlighting is legitimate up to a certain level, it will be easier to regulate and to determine when the 'bit on the side' becomes a full-time business.

Any discussion of the black economy must distinguish between undeclared personal services (as discussed here) and criminal activities associated with work, usually in connection with formal employment. This criminal black economy is well described by Gerald Mars.[6] The devices by which more and more people enrich themselves, usually at the expense of their employers, do not amount to a 'new world of work', but they are a distressing sign of the low esteem in which many hold the old way of work. Cheating at work is today expected, even condoned, in industries like catering, transport and retailing, while the personal use of office telephones, stationery and secretaries is an accepted perk of every manager. Compared with many of the practices of formal employment, moonlighting is a paragon of honesty and integrity, at least in customer relations.

The mauve economy

The respectable cousin of the black economy is what Helen Chappell calls the 'mauve economy'.[7] As the employment economy becomes more inaccessible to more and more people, they start to create their own forms of work. These are usually services, some of them bizarre (singing telegrams, dating agencies for single parents), some of them mundane but done from home (book-keeping, part-time typing, printing, cooking) – the kinds of service that people are prepared to pay to have provided for them if they are in employment and have money.

These businesses are usually too small to need to be registered for VAT. They are the 'miscellaneous services' of the government statistics – if they ever get into the statistics. A Labour Survey in 1982 found 750,000 more people working in Britain than previously estimated,[8] most of them in the tiny businesses of the mauve economy. They are not necessarily breaking any law or regulation because tax on the self-employed is collected in arrears, after the first year of trading, but such businesses do not keep PAYE accounts or inquire whether or not the people whom they hire to help them are on the dole. They are therefore hard to spot and to count. They are

BOX 3.4   A PILLAR OF THE MAUVE ECONOMY?

The Memory Bank was started by Roger Jameson and Stephen Pedvin in Roger's home in Islington. Roger is a songwriter and Stephen an 'alternative accountant' who gives his services free to housing corporations and yoga groups. They came together to provide an unusual personal service – a reminder service.

The Memory Bank charges £14.90 for an annual subscription to the service, which covers the first ten reminder calls. Clients are reminded of their appointments – doctor or family planning, birthdays, business meetings, parties – the day before.

They hope the business will be self-financing and will need few overheads. They are buying a small home computer to cope when the workload increases, and plan all kinds of advertising ploys – leaflets, tube posters, questionnaires, hiring the sides of buses. They foresee a network of housewives and house-husbands taking on the work on their behalf. 'People expect more and more in the way of information today,' says Steve, 'and those in work can afford to buy it.'

Helen Chappell
'The Mauve Economy', *New Society*, 28 July 1983

betwixt and between the formal and the informal economies, a respectable fringe.

Six million people now work for Britain's 1¼ million small firms, and the number of self-employed people is growing by 5 per cent p.a. The Government encourages the official end of it – over seventy measures were introduced by the Conservative Government in 1979–83 to support small firms, and the Enterprise Allowance Scheme run by the Manpower Services Commission pays a salary to would-be entrepreneurs among the unemployed, most of them going into 'miscellaneous services'.

Some people have been busy turning their hobbies into businesses or their idle skills into a paying part-time service. The pate made for home entertaining, for example, becomes demanded at local charity sales and ends up being made under contract for the local delicatessen. The list below gives some of the many possibilities for a home-based business:

carpentry	taxi service
dressmaking	photography
tutoring	typing
book-keeping	writing
seedlings	journalism
job printing	toy-making
car maintenance	pottery
tree pruning	plumbing
vegetables	housing consultancy
horticultural advice	painting
computer programming	translating
interior decorating	reading scripts
indexing	picture framing

The mauve economy is undoubtedly flourishing in pockets around the country as more and more people get nudged towards self-employment because they cannot get on to a payroll. It is a new world of work. It is, however, still a fringe world, not the dramatic change that some enthusiasts describe. The *Economist* in July 1983 sounded a note of warning, quoting an American software expert: 'There's something in this country that leads people to look for security in employment. When you interview people (for a job) they often ask you about the pension scheme very early on. In the United States it is never high on people's mind.'[9] The head of Berkshire's enterprise agency says of the people coming to him with ideas: 'I certainly wouldn't describe most of them as entrepreneurial.' The *Economist* rightly points out that Britain's competitive future will depend on how many efficient firms it spawns in new industries, not on the total number of small firms in its economy or on the numbers of the part-time self-employed.

It is, however, early days. The first signs of spring are few and tender. It is certainly a change that one-third of unemployed people, according to a poll commissioned by the Economist Intelligence Unit,[10] had thought of starting their own business. Even if the new small service businesses of Britain do nothing to help her compete internationally, they are still a new world of work, a positive alternative to the 'bread and circuses' of unemployment. The tiny businesses and part-time self-employment of the mauve economy look like being a growing sub-culture. Some of these enterprises will grow into proper businesses in the formal white economy; some will die; others will remain small but crucially important to those who work in them. It is a sub-culture which needs nurturing because it does signal a change in the culture. If more and more people think of creating their own work rather than waiting for it to be created for them, the American visitor's criticism may one day seem very dated.

The grey economy

Whereas the black economy is small and illegal, the grey economy of the household and the community is perfectly legal and very large: it probably includes all of those 40 million people who are aged between 16 and 74 for part of their time. Richard Rose estimates that it may have accounted for 51 per cent of our labour hours as a population in 1975, compared with 46 per cent in the formal economy and 3 per cent in the black economy.[11] Such precision has to be suspect, but the order of magnitude is probably right.

The average adult has sixty-one hours a week of free time (not spent in formal working or in sleeping). That is much more than the average person spends at work in a job. Any or all of that time can be devoted to doing for yourself things that you could buy from other people or pay other people to do. Most of us do our own cooking, cleaning and washing. Most of us rear our own children and cut the grass in our own garden if we have one, as we do our own shopping and mending. But not everyone does. If we were all rich enough, we could pay others to do these things for us. In fact, theoretically, we could pay each other to take in our washing or employ each other's spouses as our housekeepers or odd-job people. In other words, in our domestic activities each of us is replacing formal economic activity with our own labour.

Richard Rose has again calculated that while we spend the biggest chunk of that free time doing nothing or on pastimes like walking or talking that cost no money (twenty-one hours a week), and some of it on leisure activities like travel and pubs that do cost money (seventeen hours a week), we spend even more of it (twenty-three hours a week) on unpaid housework and productive leisure (gardening or knitting).[12] We could spend even more. Rose's table of opportunity is provided in box 3.5.

As Gershuny has pointed out most households nowadays have enough capital equipment in their kitchens and living-rooms to run a business if they want to look at it that way.[13] Alcoholic drink, for instance, is the major opportunity for substituting domestic product for a bought product, and most households already have the bulk of the equipment needed in terms of pails and containers. Many of the items in the 'high substitutability' category in box 3.5 can be met in ways which use capital-intensive equipment and make the labour element minimal and less tedious. Washing machines, lawn mowers, vacuum cleaners and microwave ovens take the chore out of much household maintenance.

Hitherto this domestic or grey economy has been the counter-weight to the official economy. When times are good we have spent more in the official economy and used more of our free time on pastimes or leisure consumption. When times are bad we do more for ourselves. This has therefore accentuated the height and the depth of the economic waves. Will it continue to work in this way in the future?

There are some reasons for thinking that the grey economy will continue to grow rather than to move up and down. The main reason is that the amount of free time in each lifetime is bound to increase for the reasons given in chapter 2, while at the same time lifetime earnings after tax may well decrease (because we shall have shorter working lives and probably a higher rate of tax). Therefore even if we wanted to we could not spend a much greater proportion of our time on leisure consumption which has to be paid for. The incentive to save expenditure by doing things oneself will be high.

Secondly, fashion may be changing. In a materialist society the more you buy visibly in the official economy, the higher your status. In a society where capability and self-sufficiency are valued, it is counted better to build your own bookcase than to buy it, to cook in your own home than to eat in a restaurant. The ease and sophistication of the new kinds of capital equipment for households helps to make capability easier as well as more fashionable. In Jonathan Gershuny's phrase, we may be seeing the dawn of the 'self-service society'. To put it another way, 'success' may come to be seen as one's degree of independence. Paying other people to do or make all you need actually makes you dependent on them. Fashion used to call that wealth and success, but fashion can change. Perhaps

BOX 3.5   THE OPPORTUNITIES OF THE DOMESTIC ECONOMY

The distribution of household expenditure in the official economy
by likely opportunity to substitute in the domestic economy

	£ (bn)	%
High substitutability		
Alcoholic drink	11.4	
Catering and accommodation	7.6	
Household maintenance	5.0	
Miscellaneous recreational goods	4.3	
Other services	5.3	
Domestic service	0.5	
Total	£34.1	(23%)
Medium to low substitutability		
Food	24.1	
Clothing	10.1	
Travel	5.2	
Maintaining cars	3.7	
Entertainment services (TV purchase;	3.0	
admissions, etc.)	2.6	
Other miscellaneous goods	3.0	
Expenditure abroad		
Total	£51.7	(34%)
Nil substitutability		
Housing rent, rates	19.2	
Durable goods	12.2	
Fuel and light	7.8	
Tobacco	5.5	
Running a car (petrol, licence)	6.5	
Other household goods	3.9	
Chemists' goods	2.3	
Post, telephone	2.5	
Insurance	2.2	
Books, newspapers	2.2	
Betting and gaming	1.3	
Total	£65.6	(43%)

Source: Compiled by Richard Rose from Central Statistical Office, *National
Income and Expenditure* (1982), table 4.6, consumer's expenditure at
current prices.

it should in a democratic society, for materialist wealth has meant
that while some could be rich, many were bound to be poor by
comparison. If wealth means self-sufficiency, then everyone could,
in principle, be rich and successful.

Thirdly, there is mankind's need to express itself in work. If jobs are not guaranteed, it may be more sensible to express some of oneself in domestic work. Betty Friedan quotes an American automobile worker: 'There's no security in a job. The dollar's not worth enough any more to live your life for it. I'll work three days a week at the garage and my wife will go back to nursing nights, and between us we'll take care of the kids.'[14] There is anecdotal evidence from parts of Britain that young married men, disillusioned with the boredom and insecurity of the job, are turning to the home for much of their satisfaction, while their wives flee the drudgery of household and childcare in the job market. The grey economy is of particular importance for unemployed people. Kilpatrick and Trew found that unemployed men in Belfast were likely to spend an additional sixteen hours a week in domestic work, partly in childcare, partly in do-it-yourself activities.[15] If unemployment increases, as chapter 2 suggests it may, then the grey economy seems bound to grow as more of the unemployed find work for themselves, and a way of saving money, in the grey economy.

Is a growing grey economy good or bad? For individuals it is mostly good. It provides a cushion against the vicissitudes of the job and the changing fortunes of the formal economy; it is an outlet for creative energies and a way of defining oneself apart from the job. Women may not agree. To women the grey economy has traditionally been a drab and boring prison, made more lonely by capital equipment and convenience foods which make one so self-sufficient that one has no need of human contact. It seems to be that as *part* of one's life the grey economy is good news; as the *whole* of life, it is bad.

For governments a growing grey economy is likely to feel like bad news. It means less visible economic activity, therefore less taxable money in circulation and fewer jobs. Traditionally governments have been pleased to see that prosperity, industrialization and urbanization have conspired to destroy the self-sufficiency of the home because that has forced more economic activity into the open. An inter-city high-rise apartment has little capacity for substituting for anything in the formal economy. The result has been to lever ever upwards the slope of apparent economic growth. If you live in a tiny apartment, there is an ever-present urge to go out, and every time you go out you spend money or make money; either way you fuel the economy. This is, however, a dangerous route for us to follow to extremes. The citizenry could become totally dependent on income to survive, which would put an obligation on the state to ensure that everyone's income is above a certain minimum if he or she is to have access to the things that society has to offer. It is no use preaching the virtues of self-reliance to people who have been taught to think that success is being able to buy everything they need and who no longer have the space, the tools or the skills to make what they need. In theory there is no limit to the amount of money that can be sloshing around in society keeping everyone busy buying and making. In practice no government has found a way to add to the pool without causing inflation, reducing the value of money and so getting back to where it started. A society that pays for everything it wants or needs can look rich, but it works only if everyone has an adequate income – something devoutly to be wished, maybe, but hard to deliver.

A growing grey economy may not, therefore, be totally unwelcome to governments, in that it will take a little of the pressure off them to provide everything for everyone. Indeed, the surprising lack of aggression by the unemployed in all Western countries to date may be due in part to the cushion of the grey economy. We have, however, to remember that no one in an industrialized country can live without some dealings with the formal economy. Everyone needs *some* money. The grey economy is not a lifeboat, only a piece of wood in a strong sea.

NEW MEANINGS OF WORK

The black, mauve, and grey economies represent new or resurrected opportunities for work outside the job. They underline the dawning realization that there is more than one form of work. We can best describe the differences by thinking of three different forms of work:

(1) job work, which is the paid job, including full-time self-employment;

(2) marginal work, which covers the work we do 'on the side' for extra earnings, which should be, but sometimes are not, declared. (You might call it 'pocket-money work'.)

(3) gift work, which includes all the work we do for free in the grey economy and in voluntary work.

If work is redefined to include the extra categories of marginal and gift work, you get a much truer picture of what people do in this country.

Consider these facts. Fewer than 20 million people out of Britain's 56 million have full-time jobs. A further 13 million are adults of working age, of whom 4 million might describe themselves as not working and wanting jobs. The other 9 million may work part-time work (4 million do); they may do voluntary work (18 million people do some of it); they may do something 'on the side' in the black economy (probably 1 million do some of that); or they may be wholly or partly occupied in the grey economy of the household (4 million women are probably in this category). Few, if any, of these 9 million people would think of themselves as 'not working'. By restricting the word 'work' to mean a 'proper job' we have, as it were, disenfranchised at least 9 million people.

No one would want to argue that marginal or gift work is an adequate substitute for job work if jobs are what you are looking for. The suggestion to an unemployed person that there is unlimited scope for voluntary work will often be heard as an insult ('We can't pay you, but we are prepared to use you if you cost us nothing'), for gift work has to be given, not demanded. A gift that becomes obligatory is a tax.

BOX 3.6   WHAT PEOPLE DID IN BRITAIN IN 1981

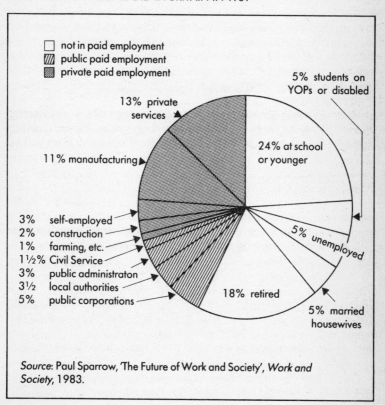

□ not in paid employment
▨ public paid employment
▩ private paid employment

5% students on YOPs or disabled

13% private services

24% at school or younger

11% manaufacturing

3%   self-employed
2%   construction
1%   farming, etc.
1½% Civil Service
3%   public administraton
3½   local authorities
5%   public corporations

5% unemployed

18% retired

5% married housewives

Source: Paul Sparrow, 'The Future of Work and Society', *Work and Society*, 1983.

The point of the threefold definition of work is to give us a more rounded picture of work in society. Jobs are important, but they are not the only way in which a man or a woman contributes to society, finds an identity, meets friends or makes money, which are usually given as the reasons for having a job. A survey in the *Guardian* newspaper in December 1981 asked what were the main elements in job satisfaction. Top of the list came personal freedom, respect of colleagues, learning something new, challenge, completing a project, helping other people. Twenty-fourth on the list was money; seventeenth was security. The respondents may have deluded themselves or the *Guardian*, but what they seemed to be saying was that work provided them with the opportunity to fulfil themselves, to grow and to work with and for others. That is another way of saying that work is essential to the full expression of our humanity.

But does that work have to be a *job*? The perfect job provides us with money and security as well as the six first items listed above. But how many jobs are perfect? How many will be as perfect under the new technology? I once watched a man watching a machine putting mints into a pack. He was there in case the machine missed a mint. It never did. What kind of work is that for humankind?

Not all jobs will be as dehumanized. Indeed, the new technology can turn the secretary into an information manager, the ticket clerk into a personal travel agency and the car mechanic into an automobile doctor. It works both ways. Many people, however, seem to be hedging their bets these days. They do not rely on the job alone to give them the full package of satisfaction which people rightly expect from work. They are looking to get different bits of the package from different bits of work.

Work, for instance, can be viewed as a commodity or a vocation. If it is a commodity, the worker is selling his or her time or skill for a price. He or she is entitled to look for as big a price as possible, while the buyer, the employer, will pay as little as is practicable. It is a bargaining process, ritualized in the collective bargaining of trade unions and employers. Where work is a vocation it is done because you want to do that work; it is your interest, your profession, your commitment, maybe even your passion. Through that work you contribute of yourself and are more likely to win the kind of respect you really appreciate from colleagues and friends, the kinds of satisfaction that mean something. Perhaps boring, unsocial work of a purely commodity kind deserves to be rewarded more than work which brings its own intrinsic satisfaction. Maybe garbage collectors *should* be paid more than dons (see box 3.7).

Lucky are they for whom the vocation and commodity views come together in their job. Professionals, managers, skilled craftsmen and

BOX 3.7   HOW DO YOU PRICE THE COMMODITY?

The Vice-Chancellor was complaining that after graduation the salaries of his students were still lower than they would be if they had gone straight on to the assembly line in the local car plant.
'Is that not as it should be?' replied a foreman at that plant. Your graduates go out into a world where they will work in offices with carpets, with time to talk, doing interesting jobs in civilized surroundings. My people do jobs just as important but in a noisy, dirty place, jobs that are monotonous and tiring even if they are important. Is it not fair that they be paid more by way of compensation?

technicians get well rewarded for doing what they enjoy doing. ('Happiness,' someone once said, 'is being paid to do what you would do anyway.') Other people need different bits of work to satisfy their different needs. Job work may provide the money, but gift work or marginal work provides the interest. The complaint of the older worker that today's youth has little pride or interest in the job, that it takes the money and runs, is one indication of how instrumental the job has become to many. The shortage of many types of worker in a decade of high unemployment suggests that work as a commodity is still being priced too cheaply if people can manage to get money elsewhere. Nasty work will become better paid, with shorter hours, as people sell their commodity work more dearly, dealing wherever they can.

Work, in some form, is critical to individuals. It is, apart from anything else, a principal *structure for mattering*. We all need to feel that we matter, that we can contribute, that we are missed in our absence, that we are respected and liked. If jobs are scarce, or not much good when we get them, we must and will find other forms of work. In the past we were perhaps obsessively preoccupied with the job. Those who didn't have jobs, like housewives or the elderly, were seen in some ways as extras in the play of life, the chorus rather than the principals. No wonder that women have been eager to get in on the act. We badly needed, therefore, to recognize that work did not only happen in jobs. That sort of recognition is beginning to come through, partly because more and more men are finding that work has many faces, a truth that women knew long ago.

The survey of people doing volunteer work reported in box 3.8 indicates that this form of work provides many people with the satisfaction of the more vocational side of work. It also indicates that gift work is not confined to the females, the middle-class and the

BOX 3.8   VOLUNTARY WORK: WHO DOES IT AND WHY?

The popular image is that a typical volunteer is a woman with no
work and no children, middle-aged and middle-class. The truth
comes from a survey conducted by the Volunteer Centre in 1981.

Age	*% of sample*
18–24	14
25–34	19
35–44	16
45–54	17
55–64	16
65+	18

Sex	
Male	46
Female	54

Marital status	
Single	32
Married	67

Occupational group	
Professional/managerial	18
Intermediate	17
Skilled manual	36
Semi-skilled	16
Unskilled	5
Forces/unclassifiable/housewife	7

Satisfactions	
*I do it because I really enjoy it	51
*The satisfaction of seeing the results	50
*I meet people and make friends	38
It is part of my religious belief	31
*It gives me a sense of achievement	30
Because others are less fortunate than I	30
*It gets me 'out of myself'	26
*I like to feel I'm needed	25
It broadens my experience of life	25
*It gives me a chance to do things I'm good at	23
It makes me feel less selfish	22
Because somebody has to	19
*It gives me a position in the neighbourhood	8

Note: *Commonly stated reasons for job work in other surveys.

elderly. Almost half of the cross-section of people interviewed in the
survey had done some voluntary work in the preceding year. Gift
work is part of everyone's life. No statistics are available for marginal

work, but, as our discussion of the black economy demonstrated, almost half of those employed *could* carry their skills over into marginal work, and everybody *could* turn his or her hobby or passion into a mini-business. The mauve economy shows that more and more people are beginning to do this. Work is being redefined not by legislation but by the actions of people who feel, quite rightly, that life needs work of some sort to be a proper life. If there are not enough jobs, they will slowly, creatingly and painfully, in many cases, create other forms of work for themselves. We should be glad, not censorious.

NEW PATTERNS OF WORK

The short supply of jobs is going to put continuing pressure on the 100,000-hour job. If something is in short supply, there are two possible responses: one is to hoard it; the other is to share it out more thinly. We can today observe pressures to do both of those things.

On the one hand, all sorts of people and institutions point out that four men working for five days could be replaced by five men working for four days. It is not quite as simple as that, as we shall see, but the principle of work-sharing is increasingly widely supported. On the other hand, those who are already in those jobs are understandably reluctant to give away any of what they have got without proper compensation. Four days instead of five is fine, they argue, as long as we are paid for five. The commodity principle is at work. That response leaves nothing in the wage bill for the extra people, implying that money would have to come from increased productivity, but if productivity increased by 20 per cent, the extra workers might not be needed at all.

In the long run the social pressures on the hoarders will be too great. The 100,000-hour job is already crumbling down to 75,000 hours in many offices, where a thirty-five hour week is the norm. The likelihood is that the 50,000-hour job will be commonplace by the early 1990s. In theory this ought to be double the number of jobs available, but in practice, as we shall see, the main result will be to increase the leisure or discretionary time of the ordinary individual. It will radically transform our patterns of work over our lifetime. From 100,000 hours to 50,000 hours with little increase in jobs? The numbers and the conclusion are startling. Let us see how it could happen.

> 100,000 hours = (for example) nearly 47 hours per week
> (including overtime) for 47 and a bit weeks
> for 47 years

50,000 hours = (for example) 32 hours per week for 45 weeks
for 35 years

A lifetime of 50,000 hours' work would mean shorter working weeks, shorter working years and shorter working lives. Each has its supporters and its opponents. All will happen, slowly but surely. None will have much effect on the number of jobs. We need to look at each in turn to see why this is so.

Shorter working weeks

The French Government has announced its intention of reducing the working week to thirty-five hours, already the effective working week in most offices. It could obviously go further. There have been several forecasts of what would happen to employment if a universal thirty-five hour week were adopted in all firms. At one extreme the theoretical answer could be an extra 2 million jobs, but the Department of Employment considered that an estimate of 100,000 to 50,000, with 7 per cent added to labour costs, was more realistic.[16] The lower figure in that forecast seems more probable if one considers what has actually happened. A study of twelve cases of shorter working time produced, in the short term, little effect on recruitment.[17]

The first step, however, could be to reduce overtime. Compared with other countries, Britain works a lot of overtime, particularly among manual workers:

	45 hours per week or more (% of manual workers)	48 hours per week or more (% of manual workers)
UK	27	20
France	23	9
Germany	15	9
Italy	12	9
Netherlands	10	7

Source: 'Work-Sharing and Unemployed', a discussion paper of the Institute of Personnel Management, May 1983.

One estimate has suggested that overtime working in manufacturing represents half a million jobs.[18] Another has suggested that a curb even on excessive overtime (eight hours a week) would create another 100,000 jobs.[19]

The French have recently introduced an annual overtime maximum of 130 hours a year (Austria has sixty-five). The British so far have been reluctant to intervene by fixing a legal maximum. Overtime is an invitation to hoard work. It is convenient for employers, who like the flexibility, and it boosts the real wages of the workers. Surveys show, however, that while they may like the money, long hours result in dissatisfaction among workers.

The trade unions want shorter working weeks but for the same money. They want reduced overtime but without loss of income. Increased productivity might pay for more money and fewer hours. It is most unlikely to pay for more money, less time and more people. The most likely forecast, therefore, is that the official working week will gradually shorten to thirty-five hours and that overtime will be progressively reduced by industrial agreement rather than by law, and as industry becomes more and more automated, but without many additional full-time jobs. The hours worked in industry will continue to be greater than those worked in the services, but the proportion of people in services will keep the average down.

But our calculation shows thirty hours a week, not thirty-five. That is because increasing numbers of people are likely to be working part-time *at some stage in their lives*. The number of part-time jobs is increasing. By the mid-1990s the part-time workforce is likely to be close to 30 per cent of the total labour force. Averaged out over the whole working population or over one person's lifetime, this will bring down the average working week to thirty hours. *This is important.*

If a shorter working week is going to have any significant impact on the numbers of jobs available, it will be because more part-time jobs are available. If a firm cannot make up its labour shortfall by overtime, it will, if it possibly can, take on short-term, part-time workers rather than increase its full-time workforce, or it will contract out the work to the part-time self-employed. At present in the UK if a person works less than sixteen hours per week, neither employer nor employee pays National Insurance. The temptation is obvious – so obvious that this regulation seems bound to be changed.

There are two other possible consequences of a reduced working week. First, more organizations will start to think in terms of shifts. It makes little sense to keep your plant, or your shop, or your office standing idle for twenty out of the fifty hours it could be open just because your labour contracts specify thirty hours. New shift agreements in the services may well be paid for in time rather than money, with one person working ten hours a day from Monday through Wednesday and another from Thursday through Saturday (thirty hours each) whenever the type of work allows it. In practical terms this will be the job-splitting of the future, but it will be called 'shifts' or 'rotation'. Second, as the organization of work becomes more flexible and more individualized (for reasons explained in the next chapter), organizations may find it not only possible but desirable to make individual contracts for different quantities of hours to suit the individual and the demands of the work. We might call it 'giant flexitime', but actually it will be individualized part-time work.

BOX 3.9   TIME, NOT MONEY, AS PROFESSIONAL PAY

Stephen Bragg, chairman of the Cambridge Health Authority, has a radical proposal for solving the staffing problems of hospital doctors.

At present four times as many consultants are needed as housemen (junior doctors) to allow all who want to be promoted, and are qualified, to become consultants. But at the same time twice as many housemen are needed as consultants if each 'firm' of consultants is going to have one houseman on duty at all hours of the day or night, as required by hospital rules.

The situation seems intractable unless, says Stephen Bragg, you encourage the more senior doctors to work fewer hours for the same salaries as their juniors. House officers could work a six-day week on a shift system for a regular salary; registrars, being rather more senior and experienced, four and a half days for the same salary; and consultants one and a half days for the same salary. This would allow the number of consultants to be increased by 300 per cent, but there would still be one houseman to every full-time equivalent consultant (roughly the ratio at the moment).

Consultants could choose to make more money in private practice in the remaining three and a half working days per week or to have more free or private time. Jobs would have been increased, work and money shared, portfolios of work adjusted.

Could it work in other professions?

Both of these developments, by making more efficient use of the capital assets of the organization, could produce more but smaller jobs because the added productivity would pay for them. We may therefore expect to see this happening wherever more intensive use of plant will pay off in the market. It already happens to a degree in retailing, in factories with seasonal order peaks, in hotels and in catering businesses. It could spread as the service sector spreads and as manufacturing businesses effectively become service organizations with an automated production function (as the next chapter will explain).

Shorter working years

At present everyone at work has a minimum of three weeks paid holiday plus eight public holidays. Over 75 per cent already are entitled to four weeks, and the percentage grows each year. There seems to be every likelihood that this will increase to five or six weeks for nearly everyone in employment by the 1990s. That will mean a forty-five-week year. The gradual reduction in working time is most unlikely to result in any increase in the number of jobs. It is only a formalization of the absenteeism currently experienced in all organizations in all parts of Europe and America.

Could the working year get any shorter? Suggestions have been made: sabbaticals, extended training, even a legal requirement that everyone should have a year off on full pay every ten years. There will almost certainly be a trend towards more education during a working life, and for some this may take the form of an extended course, the equivalent of a sabbatical. There is, however, no evidence that anyone really wants as long as a year or even six months off work, even on full pay. When the sabbatical was built into steelworkers' agreements in the United States it was found that most sought alternative work during the period. In other words, even if a year's pay for no work were agreed and accepted, it is unlikely that new jobs would be created; all that would happen would be a shift in labour.

The evidence is that people would prefer to take two short holidays, if they can, rather than one long one.[20] Short holidays make for less disruption to organizations than long breaks and can therefore be accommodated without hiring extra staff. One estimate suggests that an immediate additional week's holiday for all employees would result in only an extra 25,000 jobs.[21]

The working year, we may assume therefore, will shrink without creating any more jobs.

Shorter working lives

Most men still work until the age of 65, most women until 60, but the percentage of people that do so is falling fast. From 16 to 60 is forty-four years; it is forty-nine from 16 to 65. If we are to see any significant sharing out of work, it will come from a shortening of the working life.

The existing workforce can make good a missing hour or two or an extra week in the year's holidays – indeed, it is in its interest to do so because that allows it to maintain its levels of earnings. Shorter working lives, however, actually remove people from the payroll. If working lives were progressively cut back by twelve years, workforces would shrink by 25 per cent, if no replacements were found. Just to keep up existing levels of production firms would have to find more people.

In fact, working lives have already been pruned but mostly at the front end. In the 1980s it is increasingly difficult to enter the workforce before one is 20. The gap is being filled by education, mixed with enforced leisure. More and more men would like to leave the workforce at 60 or before if they could afford to. Some can and do. The managerial classes in the bigger firms join as graduates at 21 or 22 and leave in their late fifties on an early pension (that is, they work for between thirty-five and thirty-seven years).

If the state retirement age for men were reduced to 60, it is estimated that 400,000 males might opt to retire; they would be replaced by about 266,000 new recruits at an additional cost of £2.5 billion p.a. to the state and the same occupational pensions schemes.[22] The cost of such a universal measure seems unacceptably high. To many it also seems to be a restriction on people's right to continue to work if they so choose, and indeed the official thinking in government is that the retirement age of women should be raised so that it is closer to that of men, perhaps letting them eventually meet half-way. Nevertheless, West Germany has announced its intention of fixing 59 as the official retirement age.

A more likely alternative is a move towards flexible retirement, as has been tried in some European countries. In Belgium, for instance, pensions are available up to five years before retirement at levels actuarially reduced by 5 per cent. Reduced pensions have already been proposed by a Select Committee of the House of Commons, although they could not be introduced until the mid-1990s. In Holland the state pays the individual half of what he or she loses by retiring early, expecting the employer to pay the other half. In Britain the Job Release Scheme was intended to ease out older workers in favour of new recruits, but the low level of allowances and the requirements that the departing employee be replaced directly by someone on the employment register has limited its impact.

Dramatic measures and changes are unlikely. They would be too expensive. We are, however, likely to see a continued erosion of the working life at both ends:

(1) pension schemes (state and occupational) will become more flexible, allowing for increasingly early retirement;

(2) part-time working for older workers will be introduced, without loss of pension rights;

(3) the years between the ages of 16 and 19 will increasingly be seen as years of education, training and work experience, postponing entry to the workforce until 20;

(4) individuals will be allowed to take periods out of work – to raise children, to go to college – without punitive loss of pension rights;

(5) personal and portable pension schemes will become more common among the professional and managerial classes, making it easier to opt for early retirement or self-employment.

Later chapters will argue that these measures will not be enough, that more will need to be done for more people 'beyond employment', but nevertheless by the 1990s it will be unusual to find anyone expecting to have a job for more than forty years, while the thirty-five-year job will be a reality for many.

This steady erosion of working life is not going to create huge numbers of new jobs. What it will do is to redefine the normal working life and therefore the 'working population' in the sense of those who are eligible and anxious for a job. If most of those over 60 and most of those under 20 were excluded from the number of potential job-seekers, unemployment would fall by over 1 million, to which reduction could be added the 250,000 or so new recruits taken on directly or indirectly in their place.

The 50,000-hour job is therefore not a panacea, not an imaginative national work-sharing plan, but an economic compromise. It may, it is true, produce an extra half-million or so jobs, many of them part-time. It may also, eventually, cut the numbers of those wanting jobs by 1 million; and a 1½ million cut in unemployment should, on the face of it, be good news, but it will have come about largely because employers want lower costs and thus higher productivity, while workers want higher wages for less time. Higher rates per hour or per year can be afforded only if there are fewer hours and years to pay for and the same or better output. In other words, the savings that result from producing the same goods with fewer man or woman hours are shared between the employees (in higher rates) and the employers (in higher productivity).

Looked at the other way round, however, the work-sharing

element is more obvious. The same result could be obtained by keeping a 100,000-hour workforce but cutting its numbers in half and paying more to those that remain. There are signs that this is the preferred solution of some firms, some institutions and some closed professions who have cut their recruitment to a favoured few, raising the drawbridge, as it were, on youth. They run the risk of becoming increasingly middle-aged institutions and increasingly expensive, since on the whole the old are more expensive than the young as well as being less energetic and less creative. If the armed services had responded in this way after the end of National Service and had retained the concept of lifetime careers for a select few, the Falklands campaign would have been fought by platoons and squadrons of 50-year-olds. Instead they opted for a range of short-service commissions, thereby increasing the flow but not the cost. In other words, without the halting progression towards the 50,000-hour job, things could be even more difficult.

The effect of the move towards the 50,000-hour job will be to increase the flow of people through much the same total quantities of smaller jobs. Because increased productivity may permit the same or even higher rates of pay for fewer hours, in the short term there will be more satisfaction on the part of those in the jobs. The pain will come later on because the lifetime earnings from 50,000 hours can never be equivalent to those from 100,000, even if they are considerably better than half.

But who has ever paused to calculate his or her potential lifetime earnings or realized that he or she is expected to work 80,000 or 100,000 hours? The 50,000-hour job is, in the end, a way of sharing work, but it is also, in the end, a way of sharing income. This is the most pressing reason why the job cannot be the whole of life, even if there were enough of these smaller jobs for all. That is why the other forms of work are so crucial because they either provide extra money, save money or substitute for money. These new meanings of work are not alternatives to the job; they are essential companions to it.

TWO OPTIONS

These ideas add up to one view of the future of work, but there is a very different view. Here they are for comparison.

Option 1

This option is positive. The 50,000-hour job, in this option, will share out the jobs but in a different way. Because the working population will gradually be redefined, that population (between the ages of 20 and 58) will be much closer in numbers to the jobs that are available, particularly if some of those jobs are multiplied by more shifts or part-time work. Everyone between the ages of 20 and 58 will have a much better chance of 50,000 hours of a job. Smaller jobs and fewer people wanting them will produce a better match.

Redefining the working population will get rid of unemployment as a statistic. It will not, however, give the 19-year-old or the 59-year-old anything to do or much to live on. That is why the new worlds of work become crucially important in the scenario. A 50,000-hour job will dramatically change the place of the job in life. the job will become very much a part of life, even a stage in life, rather than the whole of life. There used to be a saying, 'Born a man, died a baker', implying that the job provided people with their whole identity, and all their money lasted until death. There used, too, to be an American business whose pension fund averaged payments of only three months per individual, so quickly did employees die after retiring. It will not be so again. The average person can expect fifteen or twenty years of active life after the end of the job, and even during the job he or she may well have three days off a week and seven weeks off a year.

What will we do in all this discretionary time? Watch television, walk the dog and go to the pub? Yes, maybe, but will that be all? What we know of mankind suggests that for most of us it will not be all. Work is the measure of man – good work, that is, work which allows us to express ourselves, which provides an outlet for our creative energies, our ability to work with and relate to others, work which is under our control, not in control of us.

If we cannot find that sort of good work in the job, we shall create it for ourselves in the alternative worlds of work now that we have the time and the energy and the basic wherewithal provided by the 50,000-hour job. While some will still work all the hours they can, combining commodity and vocation in their job, many more will end up with 'portfolios' of work, a mixture of job work, marginal work and gift work.

The mix in the portfolio will change as we go through life, but maybe the changes will be less abrupt than they are today. Job work will continue to dominate our lives from mid-twenties to the end of our forties, with intervals for child-rearing and education, but gift work and marginal work may be increasingly important and may predominate in our fifties and sixties. It will all be work, but we may get different satisfactions from different bits of it. The job will no longer have to carry all our needs, hopes and ambitions. All work will be respectable. There will be enough paid work to provide for each of us, in 50,000 hours, a lot (although not all) of what we need to support us financially for the whole of our lives and enough of the other sorts of work to keep us active, busy and involved. We are all portfolio men or portfolio women. It may be that our job will provide money but no status, whereas our gift work will give us our place in the community. The portfolio does not need to be equal, only balanced. We could be street cleaners for (part of) the night, writers by day.

According to this scenario, the grey economy will grow; gift work will grow and be accorded status and respect. The black and mauve economies – marginal work – will be seen as the reservoirs of new entrepreneurial talent and tolerated, even licensed, up to a certain point, after which they will be regarded as part of the formal, white, economy. Nevertheless, there are problems to be dealt with, naturally, even in this option. How does one collect enough money from a 50,000 hour job to pay for a lifetime? How can we regulate the black economy? But these are technical issues if it is accepted that work is more than a job, that portfolios of work are the right thing and that 50,000-hour jobs are the norm. Work will then be redefined.

Option 2

Option 2 is more pessimistic, even alarmist. James Bellini, for instance, sees us moving into a non-manufacturing society in which industrial collapse is so fast and so widespread that the entire social fabric breaks down.[23] He envisages small, closed worlds where information is all-important and is tightly controlled by those with power. Around these small pockets of high technology with their experts will be a vast backwater economy where 'unemployment, menial work, crafts, moonlighting, barter and brigandry are standard features of everyday life'. Social groups will be sorted out into a 'hierarchy of status and economic roles that bears the hallmark of the old feudal order'.

At the top of this hierarchy will come the knowledge people; the professionals, the inventors and designers of machines and systems, the managers who control those systems and the politicians who regulate them. To these people the 50,000-hour job will have little meaning, for their talents and skills will be in demand, and they will find that the job meets most of their needs, with respect to money, security, satisfaction and identity.

The second layer comprises the modern equivalents of the hewers of wood and drawers of water. These are the machine minders and watchers of the new technology and those who lend their hands and legs to the service industries in food and drink businesses, in retailing and transport and in many of the personal services. They will charge highly for their labour, their commodity, and will cling to the job as long as they can.

Squeezed out may be the middle managers, the skilled workers and foremen who make up so much of our workforce today. Some of them may discover in themselves the entrepreneurial talent to start businesses in the personal services sector, but many will be driven to eking out a living on the fringes of society, in the black grey and mauve economies. Work to them will be a necessity and a job a privilege. Downgraded in the new social hierarchy, they will be a bitter, disillusioned and fomenting element in society. Maureen Duffy, in her novel *Gor-Saga*,[24] explores the kind of world it might be, with the 'nons' excluded from work and privileges and passports needed for places like Hampstead.

In this scenario the 50,000-hour job may indeed be statistically true as an average, but it will be an average of many widely differing jobs, from 100,000 hours down to 10,000. Those who have jobs will cling on to them. Those who are squeezed out will find it hard to get back in. The grey economy will be merely a synonym for unemployment, the black a synonym for crime.

It is a plausible, if unpleasing, view of the way things could go. We must therefore do all we can to make the first option come true. Left to themselves, things might not go that way. A free market can become a free-for-all, a jungle, and the route to the second option.

4

Reorganizing Work

New ways of work, new technologies and new attitudes to work are stirring a slow revolution in the ways in which work is organized in factories, shops, offices and banks. The mile-long factory or the forty-storey office – sheds for people armies – will one day soon look like memorials to an old civilisation. No longer do you have to gather armies of people together in one place to get any quantity of work done. Automation, cleverer and cleverer machines, need platoons of people, not battalions. The communications revolution means that those platoons do not have to be beside each other to talk to each other. Electricity allows work to be powered anywhere.

That sounds too commonplace to be revolutionary, but we still imprison ourselves in organizational forms and shapes designed when none of those things were possible. If they were to be produced in any quantity, goods needed a 'labour force', even with the help of the big machines of the industrial age. Those machines multiplied the muscles of men but not their brains – they still needed to be fed and serviced and to have their output ferried from place to place. The attendant armies of workers needed to be close to each other so that they could more easily and more cheaply talk to each other and move things from one stage of work to the next. They had also to do all this as close as they could get to a source of energy (water or coal). To gain the economic advantages of mass production organizations had to be large, in one place and in an industrial town.

The literature of management theory is devoted to the business of organizing, motivating and controlling large numbers of people whose time has been bought. It is a literature which already needs urgent revision because of the new list of options that are available in the organization of work – how it is paid for, sited, controlled and grouped. The principal options are:

tools or dials?
trains or terminals?
wages or fees?

BOX 4.1 CENTRALIZATION: BOON OR BURDEN?

In the 1970s I visited a factory in Bhopal, India, which employed 20,000 people in the manufacture of turbines for electricity generation. It was, in effect, one gigantic shed stuck down in the middle of India, with a surrounding town built specially to accommodate the workers and their families.

The factory and the organization had been designed in the 1950s by AEI (later to be taken over by GEC). AEI is reputed to have said, 'If we had had the chance, we would have put all our twenty plants in Britain on to one site. At least we can now do it in India.' The result was an organization with seventeen layers in its hierarchy, bureaucratic, incestuous and with an exceptionally high ratio of administrators to operators. The world, to those in the organization, was the organization. They were totally dependent on the organization for all their needs and satisfactions.

One wonders whether, in the 1980s, anyone would ever advise anyone to put twenty plants under one roof.

The likely consequences are:

the contractual organization
the federal organization
the professional organization
'Japanese drift?'

These options and their consequences will throw up a new list of questions for organizations, for individuals and for government. We need to think about them and to choose, so that we may steer, not drift, in the currents of organizational change.

THE OPTIONS

Tools or dials?

Schumacher, of *Small is Beautiful* fame, once distinguished between tools and machines. Tools, he suggested, were the servants of mankind, extending our reach, our muscles, our capacities. Machines, on the other hand, were our masters, requiring us to work to their pace, adapt to their requirements, go to where they were. Industrialization may have produced economic benefits, but it tended to turn people into human automata.

Our thinking about organizations has certainly been dominated by the machine. Where the machines were, originally close to the sources of energy or raw materials, there should the people be. Micro-division of labour became the efficient way to run things, breaking work down into tiny separate components so that each person could concentrate on one process – a chain of humans, looking as much like a machine as possible and, it was hoped, performing as predictably. The personnel manager of an American automobile plant once proudly stated that they could take anyone off the street and train him or her to do a job on the assembly line in one and a half hours, such was the micro-division of their work. Not surprisingly, they also had a labour turnover rate of 70 per cent a year in that plant. People do not enjoy being parts of a machine, however efficient it may be.

The new micro-technologies are changing all that. The machines are being turned either into small, automated plants which need only someone to watch the dials and to call for help if they show the wrong messages, or into very sophisticated tools which, once again, are the servants of people, this time extending our brains as well as our muscles and our reach.

The combine harvester is a familiar feature of country life these days, but not so long ago there were threshing machines tended by small armies of workers ferrying stooks of corn, feeding the machine and mending its ever-breaking belts and pulleys. That small factory of the fields has become an automated and mobile plant, run by one man perched on high. A sugar refinery which used to employ 500 people now needs five, mostly to watch the dials and to supervise the deliveries. At the same time as the machines have been turning into automated mini-factories, the tools have been getting bigger and more sophisticated. A numerically controlled machine tool may not look much like a scythe or a mini-computer like a hoe, but they all extend our capacities, are under our command and can go where we want to put them, partly because they are all powered by electricity, and today electricity is everywhere, in the remote valley as much as in the industrial town. As tools get reborn, work can move back to where people want to be rather than the people having to move to where the work has to be.

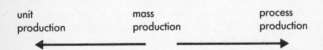

unit production   mass production   process production

The traditional categorization of ways of organizing work is a two-way split. Mass production is disappearing in factories and offices. If the operation cannot be automated, then increasingly the whole job can be done as a whole by one person or by a group; instead of everyone having a tiny bit of a whole, everyone can now have his or her own whole, aided by very sophisticated tools. The results are dramatic, in terms not only of numbers of jobs but also of the way in which work is organized.

Certainly the automated factory is on the increase, as box 4.2 demonstrates – abolishing jobs, it is true, but mostly those jobs which needed only one and a half hours' training. In any factory today the office workers will outnumber the people on the factory floor, for of all those classified as being in manufacturing the number who are actually making things is very small.

Just as significantly, the way in which those workers on the floor are organized is changing. Job shops are the current fashion. What

BOX 4.2   A 1977 FORECAST OF AUTOMATION IN THE FACTORY

By 1982 a practical adaptive-control assembly machine will be available for small component assembly.

By 1985 75 per cent of all assembly systems will use automatic inspection, and 75 per cent of all automatic inspection will use programmable control; 25 per cent of the direct labour in automobile final assembly will be replaced by programmable automation; programmable assembly techniques will be used in 25 per cent of automatic assembly operations.

By 1987 15 per cent of assembly systems will use robotic technology.

By 1988 50 per cent of the direct labour in small component assembly will be replaced by programmable automation.

By 1990 development of sensory techniques will enable robots to approximate human capability in assembly.

By 1995 50 per cent of the direct labour in automobile final assembly will be replaced by programmable automation.

These forecasts were made by 124 experts on manufacturing and assembly surveyed by the American Society of Manufacturing Engineers and the University of Michigan in 1977 and reported in *American Machinist*, 22 August 1978. The indications are that the UK is five years behind this timetable but catching up. In 1979 there were 1,014 plants in the UK with more than 1,000 workers on their payrolls. In 1983 there were fewer than 500.

R. Taylor
'Decay of the Monoliths', *Observer*, 18 September 1983

was Volvo's radical innovation ten years ago is commonplace today. Small groups of people with sophisticated tools are given responsibility for a total operation. The work comes to them, and they organize how to do it. The gang has been reborn. The original workforce was a gang – a semi-autonomous group under a gang leader who took responsibility for a piece of work. The railways and canals of Britain were built by gangs, but the advent of the machine and the factory turned the gangs into lines of assembly. Today those lines are breaking up and reforming into gangs.

It is not happening only in the factory. The assembly lines of the office (the typing pool, the ledger department) are disappearing. In the modern office there are increasingly 'gangs' of people, maybe secretaries, maybe accounting personnel, grouped around sophisticated electronic equipment, responsible for the output of a particular department. Indeed, as the equipment gets better, the gangs may dwindle to one person, the secretary becoming the 'information manager', the ticket clerk in the travel agency the 'sales agent', the insurance clerk a 'broker'.

Parts of banks, as we all know, have been automated, as has the check-out process in supermarkets and department stores. Soon even the presence of a person at a check-out desk may be unnecessary. The telephone service is becoming more and more of an automated plant. The water service has already gone that way, meaning that in many areas of the country water continues to flow even if all the labour is on strike. Even in our houses the equipment has become so sophisticated that boilers, cookers, radios, video machines and telephones turn themselves on, listen and talk and heat and cook whether there is anyone there or not. The home is still a factory but today an increasingly automatic one, with enough tools in it to equip a small business.

Automation and sophisticated tools are the news. What has not been so well understood, however, is the scope they provide to vary the organization of work. If the age of the machine has ended, if the machine has been replaced by automatic plants and super-tools, we have to build our organizations no longer around the machine but around the people, as the next two options make clear.

Trains or terminals?

In 1976 I was invited to lunch on the twenty-third floor of a newly built office skyscraper in the City of London. It was the flagship office of an international firm, and the firm was properly proud of it. Seven years later it had rented out twenty-one of the twenty-three floors because, as a spokesman said, 'What is the point of bringing people in every day and housing them on the most expensive piece of real estate in Britain and then communicating with them by telephone or computer?' So the bulk of the staff were relocated outside London, near where most of them lived.

A new insurance-broking firm chose to open its office in Henley-on-Thames, far from its traditional environment in the City of London. 'Do you not need to be in constant touch with your competitors and the markets?' I asked. 'Yes, of course we do,' they said, 'but nowadays we all do that by telephone, telex or terminals. No one goes to the bars or luncheon rooms to talk these days, so why not bicycle to work in Henley instead of squashing ourselves on to trains every morning?'

Terminals not trains. If the work is done by gangs, do the gangs have to be beside each other? Yes, if the finished output of one gang has to be transported physically to the next one. No, if the output of the gang or the individual is information. Of course, it is pleasant to know, personally, the members of the nextdoor gang, and if you know them it may improve communications and efficiency, but how much personal communication actually takes place between the 4th and the 14th floors of one of those vertical offices in the sky, or between the office and the plant? The answer seems to be very little, and probably not enough to justify the cost of bringing them all together in a very expensive place.

As organizations twist and turn to cut overheads, the economic sense of using terminals rather than trains becomes obvious if they are dealing in information. First it is the self-contained information units which get dispatched to a satellite office – the accounts department, the statisticians, forecasters and stock controllers. Then it begins to strike people that the co-ordination functions of a head office are also essentially dealing in information and might work more efficiently in a pleasanter environment, provided it was near a good railway station or an airport.

The next step is more radical. If gangs or individuals do not always need to be in the same place, do they need any place at all? Could they not work from their homes or their cars, as many salesmen do? One company puts its stock list on Prestel so that in the morning its salesmen can call it up on their own television sets for an up-to-date stock check without the need to wait for the office to open or to put in time-wasting telephone calls during the day. Writers have always worked at home. So now do more of the other information people – researchers in universities, programmers for computer firms, consultants, financial analysts and scientists. Train strikes, which affect commuters most of all, make many people aware that they do not have to be in an office all day, every day, in order to do their work. Teleconferences, once we get used to them, will often be as effective as fully fledged meetings – and much cheaper. Bleepers, cordless telephones, answering machines and telephones in our cars all make it easier to talk without touching. Terminals and telephones are taking over from the trains (which makes one wonder increasingly why British Rail does not put them on the trains).* The new technology makes possible the dispersed organization.

The dispersed organization probably means the beginning of the

BOX 4.3 NEW WAYS OF WORKING

Individuals and organizations in ten different European countries were asked about their expectations and preferences for working part-time or from home.

Do you expect the proportion of part-timers in your organization to increase in the next five years?

	% 'Yes'
Average of ten European countries	40.3
Netherlands (highest)	54.7
Spain (lowest)	12.9
UK	20.8

Do you see the possibility of more people working from home in the next decade?

	% 'Yes'
Average of ten European countries	22.4
Netherlands (highest)	34.9
Germany (lowest)	7.8
UK	16.8

If you, individually, could work from home all or part of the week, would you prefer to do so?

	% 'Yes'
Average of ten European countries	35.5
Netherlands (highest)	50.0
Sweden (lowest)	18.9
UK	40.8

If individuals want it (as in the UK), can organizations stop them?
D. Clutterbuck and R. Hill
The Re-Making of Work (McGraw-Hill, London, 1981)

*In January 1984 British Rail announced that plans for installing telephones on trains were under active consideration.

end of the gathered organization, the organizational model of the industrial age, with all its imagery of 'the works' or 'the office', the physical importance of the place and the people, the exodus from the factory gates, the mass meetings on site, the office parties, the company town and the company man. There will be many who will mourn the passing of the gathered organization, for with it will go much of the sense of community that was built into the best of the old organizations – the traditions and family feelings, the rituals and myths, friendships and companionship. There will be others who miss the power which the gathered organization gave them, be they in management or in union office. We may therefore be sure that organizations will move as slowly as they can to terminals from trains.

But the pressures of economics, unleashed by the new technological possibilities, are hard to resist. Besides, not everyone rejoiced in the gathered organization. Many it would seem (see box 4.3) would appreciate the freedom of the dispersed organization and the opportunity to work at home or part-time.

Wages or fees?

The full possibilities of the new technology and the dispersed organization do not appear until you link the physical options with the financial ones – the ways in which work is paid for. There have always been different ways of rewarding people for their time and effort (their work). Our recent preoccupation with the job has obscured some of the alternatives to the conventional wage, but these alternatives are now looking more possible and more interesting.

Wages are money paid for *time*. The individual sells his or her time. The employer then organizes the efforts of the individual during that time. Fees, on the other hand, are money paid for *services delivered*. A professional charges a fee, as does a craftsman, a plumber, a carpenter or a jobbing gardener. It is his business how he organizes his time and materials; the person who pays the fee is interested only in the quality of the service, in whether it was according to contract, on time and within budget. Fees can, of course, be paid to groups (such as a firm of solicitors), which then pay wages to their employees.

The distinction is that fees can be paid only when you make a direct link between the service delivered and the person or persons delivering it. Piece rates were an attempt to pay fees in factories. It always broke down because the measurement of the work attributable to any individual was always arbitrary. Fees imply independence, which perhaps is one reason why in Britain it has long been more respectable to be a 'professional' (a lawyer, accountant or doctor) than an employee. It is better to be your own man than someone else's. Fees, on the other hand, also suggest out-working, in which the independence of the worker has often been exploited by the organization, for independence without collaboration can equal weakness.

Ironically, many professionals are now paid wages because they work in groups or organizations, while organizations are discovering that it makes economic sense to pay fees whenever they can. Publishers, television companies and newspapers have always used fees, because they have been dealing with independent people whose services they need for some but not all of the time. They do not have to house them, service them with secretaries or coffee, or pension them. It was calculated by one multinational corporation in London in 1983 that it cost £25,000 to provide space and services for each senior executive even before they received a salary. Fees, even fees above the salary levels for equivalent work, are economic if you calculate all the costs of a wage.

Wages have other disadvantages (for employers): they are surrounded by regulations which prevent firms from abusing the time of the individual, which they have bought, and they are more and more a contract for life unless the organization goes out of business. To employ anyone these days is akin to adopting a child for life – for better or worse, he or she is yours. There are some major pressures pushing organizations towards fees rather than wages.

The fee, as we have noted, can be paid to a group rather than to an individual – subcontracting in other words. Writ large, this subcontracting shows up in Marks and Spencer's policy of contracting out all its production or Sinclair's practice of subcontracting all component manufacturing. Fees for groups are becoming standard practice. Fees for individuals are still a novelty in most industrial-/commercial organizations except when it is a long-established tradition of the trade, as in publishing.

There are signs that all this is changing, not only exotically (as in Rank Xerox's networkers: see box 4.4) but more mundanely, as smaller organizations put more and more of their fringe activities out to contract. Hire a car; do not employ your own chauffeur. Bring in a cook; do not employ one. Hire a one-man gardening service; do not employ a gardener. Contract out the cleaning; hire consultants for new product planning, an insurance broker to oversee your pension scheme, a designer to produce your company report. The language of fees is not used, but the practice is there. It will grow because it is cheaper to use spasmodic and specialist talent in this way, though it is inconvenient not to have experts around all the time. When people were cheap, convenience won over economics. Now that people are expensive, convenience has to go.

BOX 4.4 XANADU

In 1982 Rank Xerox, like many other companies, was faced with the need to reduce its overheads, in particular its central staffing costs. It could have gone down the redundancy route. It decided to do it differently.

For one group of people whose services it no longer required it organized a planned approach to independence, providing training in the setting up and running of a small business and establishing (and initially financing) a support network called Xanadu to give these new independents a basic office service and a network of friends and contacts. The company launched its ex-employees into a new way of work and life, building on the skills they had learned and practised in the organization.

With another group of people whose services Rank Xerox still required it was even more experimental. These people, several of them in top staff positions, were also set up as independents, but the company then bought back a sizeable proportion of their time and effort for a specified number of years. It equipped their new offices with the best in communications equipment, partly in order to keep them linked into the central office communications net, put them on a fee basis and linked them to themselves by terminals: a dispersed office, perhaps the office of the future. The company's savings in space and attendant overheads was such that the initial fee was often as much as the original wage, for half the time. It is a reorganization of work from which, so far, everyone seems to have benefited.

More fees mean, essentially, more self-employment. And self-employment is a rapidly growing sector of the working population (see box 4.5). It is still, however, the occupation of last resort for most people. When no one appears to want them on the payroll they will start to create their own work. The trend from wages to fees should begin to turn the page of fashion, so that it will soon be smarter to ask 'What's your business?' or 'What's your profession?' than 'What's your job?'

But self-employment is a tender plant, which needs careful nurturing in its early days. The newly self-employed need rooms and a market. It is one of the merits of the networker scheme that Rank Xerox has tried to meet both these needs for its new networkers. The millionaires of Silicon Valley all started work in their garages, but not everyone has a garage. In a tiny way some local councils are doing something about that (see box 4.5) and some agencies like the London Enterprise Agency and British Steel Industries, as well as the Industrial Estates Corporation, provide cheap factory space.

BOX 4.5   SELF-EMPLOYMENT IS A GROWTH INDUSTRY

Between 1979 and 1981 215,000 more people went into business for themselves, bringing the total in 1981 to over 2 million. It is thought that the increase is continuing at the same rate. And these are just the openly self-employed.

More than 60 per cent are in the service industries, shops, the professions, hotels and catering, with 19 per cent in the building trades and another 12 per cent in farming, forestry and fishing.

The main increase has been in insurance, banking, finance and business services, which expanded by nearly 50 per cent in the two years.

There is still scope for more ... less than 10 per cent of redundant workers in the 1970s went into self-employment, and a quarter of them had failed as businesses by 1981.

But where to do it? Will our houses need to be redesigned?

In Northumberland the English Industrial Estates Corporation, a government-financed agency set up to foster industry in declining areas, is building homes – but homes with a difference. These ones have a built-in 52-square-metre workshop intended to be the womb of a new home business.

In Essex in 1978 Asda built a pilot pair of craft homes, which were let on twenty-five-year repair leases to a repairer of airconditioning equipment and a musical instrument maker. Since then a baker, a television repairer and a blacksmith have built craft homes for themselves, working to a plan from the local council. A local developer has bought land from the council to develop six more such homes.

Guardian, 6 August 1983

Ultimately, however, customers are more essential than sheds. Self-employment, and the small businesses to which it gives birth, will really blossom when bigger businesses and other organizations like schools, hospitals and councils understand that fees are more economic than wages and start to manage a fee system responsibly, providing a market for their subcontractors.

As we shall see in the next chapter, there are consequential problems. Fees may be cheaper and more convenient for organizations, but they will not always suit the employer or the customer (or the unions).

BOX 4.6   THE LABOUR-INTENSIVE CAR?

One way in which work may shift from wages to fees is by prolonging the life of some of our major consumer goods.

The average life of a British car is now ten years. If stronger components and better anti-corrosion methods were used, the car's life could probably be extended to twenty years, but it would need a lot of maintenance and more spare parts.

Cars would be built by robots. There would be fewer of them and therefore jobs in the manufacture of cars would decline, but the maintenance activity would be labour-intensive and done largely by small garages or independent, self-employed mechanics. During the car's lifetime it would use in total more hours of labour than the 1970s car, but labour would be paid in fees rather than wages.

The same is potentially true of other consumer durables if we move to a high-quality maintenance society rather than a disposal society.

THE CONSEQUENCES

The contractual organization

Organizations may take some time to come to terms with self-employment as a way of working which is relevant to them, but there is every sign that they are waking up to the possibilities of more subcontracting on a bigger scale – paying fees to groups rather than to individuals.

It used to be that if you wanted to control something, you owned it. Organizations liked to have everyone working for them, *in their employment*. The cooks and cleaners, the chauffeurs and the messengers, even the travel agent and the financial adviser were on the payroll. The gathered organization was a total organization. The mood of the 1980s is 'back to basics'. It started as a natural response to recession, as companies cut back on paternalism and looked for ways of getting rid of all the inessentials. In the process they found that it made them not only leaner but also fitter. The most consistent recipe for organizational success is summed up by the American phrase 'Stick to the knitting'. Organizations who know what they do well and get others to do the rest do better than those who try to manage under one corporate umbrella what can be very different operations. The ways of management which suit a professional architect or engineer may not be appropriate to, or welcomed by, the cleaning staff.

Organizations have experimented with sub-contracting in many ways over time and have called it many names. Agencies have long been a way of subcontracting selling; franchises have subcontracted production and selling; and licences have subcontracted production only. Marks and Spencer, as we have seen, contract out all their production, but all manufacturing firms subcontract the manufacture of component parts, keeping to themselves design and assembly. A construction site is the most visible example of the contractual organization at work, with its lists of all contributing firms, partnerships and individuals.

At the extreme the core of an organization need contain no more than a design function, a quality-control function, a costing and estimating function and a marketing function, as well as some co-ordinating management. There are some manufacturing firms in Europe that have gone that far, and some of the newer businesses in Britain are going that route, with Sir Clive Sinclair pointing the way (see box 4.7).

The idea of the contractual organization, therefore, is not new. As we have seen, publishing, one of the oldest of businesses, has always practised it; independent television took up the concept of franchising; and Channel 4 has contracted out all its production, *à la* Marks and Spencer. Why has the idea not gone farther? Why does British Rail not sell the franchise to run a railway in each region, or even the right to serve its meals? If the brewers can sell the right to run one of their pubs, why do the retail chains still employ all the people to work in their stores?

In the state service sector, subcontracting is today a political issue and likely to continue to be so. On the face of it, it is hard to understand how a service paid for by a non-profit-making hospital could be done more cheaply and as efficiently by a profit-making outsider. The answer has to lie in the ability of small specialist groups in the service sector to run themselves more tightly. Economies of scale in services peter out very quickly because the bureaucratic burden becomes too great too soon. The cleaning operation in a hospital has to carry its share of the bureaucratic burden of the whole hospital – it has to work to the same procedures, the same rules, the same pay structures, the same conditions, whether or not these are appropriate. In large systems the highest common factor, not the lowest, prevails. In production systems large is cheaper. In service systems small is efficient.

BOX 4.7   SUBCONTRACTING

The man who used to be head gardener at ICI's Welwyn Office now runs a profitable landscape gardening business, with ICI as one of his main customers.

Enso-Gutzeit, a Finnish paper company, sold its tree-harvesting machines to the machine-operators, helped them to form their own firms and then offered them contracts to do their old jobs. Productivity has improved 30 per cent because they take that much more care of their machines.

Some former employees of Tammerneon, another Finnish company, a maker of neon lights, have converted part of its plant into an independent business. They rent all the equipment they need from Tammerneon and work on contracts. Productivity for Tammerneon has improved greatly.

On Clydeside Mr J. N. Naisly, former head of security at the UIE shipyard, reports a sharp drop in absenteeism since he became an independent subcontractor and brought most of the people working under him into a new company. They still do the same job in the same place but between the twelve of them have missed only about fifteen days in two and a half years. In the past the twenty men they used to need went absent for about 700 days a year.

Economist, 4 June 1983

An engineering firm that employs only three people and operates from a converted terrace cottage has won a £1 million order from a Turkish oil refinery for automatic loading and moving equipment. All the fabrication work would be contracted out to different manufacturers in the steel-closure-hit Scunthorpe area.

Daily Telegraph, 7 July 1983

We may therefore expect to see increasing attempts by organizations in the state service sector to subcontract ancillary services, although it is unlikely that any will go as far as television and contract out all their production or core services, but even there we may see some movement. It is inconceivable that all policing could be franchised but not inconceivable that neighbourhood watch groups could be licensed or that more security firms be allowed to operate. It is inconceivable that all education could be franchised but not inconceivable that more individual bits of education should be approved or contracted, as already happens in the Youth Training Scheme.

Whether we like it or not (and there are many who don't), the contractual organization is with us, is growing and is likely to grow faster. We would be wise to wake up to that fact because the management of contractual organizations is different from, and in many ways more difficult than, the management of employment organizations. The textbooks of management are all about employment organizations: how to organize, motivate, lead and control the people whose time you have bought. They have little to say about the management of organizations in which you can control *what* people do but not *how* they do it.

The federal organization

Small may be possible with the new technology, and the dispersed organization may cost less, but organizations are going to be reluctant to give away the advantages of size and economies of scale. They can get some of the advantages of both, as well as the benefits of independent smaller units, if they experiment with federalism. Federal organizations are composed of individual units that group themselves together for better co-ordination of their key resources (purchasing) and/or defence (marketing). In return for these advantages the individual units cede some of their rights to the centre, agreeing to pay necessary levies and to conform to certain federal laws.

We have hitherto seen more examples of federalism in nations than in organizations, probably because federalism grows most naturally from the bottom up, but co-operatives in agriculture have many of the features of federalism, as do some multinationals and conglomerates in business. We are likely to see more experiments in federalism as the way to hold the dispersed organization together.

Derek Sheane of ICI, in his analysis of federalism as a form of industrial organization, concluded: 'Industrial federalism is the appropriate form of government and management for diverse but interdependent business units facing common external threats.'[1] He need not have confined this point to business. It is true of all organizations because, as Sheane points, only federalism can give unity to diversity. Switzerland, Yugoslavia and India all contain very different subcultures which need, and want, to maintain their differences. Nevertheless, the subcultures also have to be interdependent because they are too small to go it alone in the face of all possible external threats. Federalism grows out of the three pressures of *diversity*, *interdependence* and *external threat*. These are precisely the pressures which the new dispersed organization will face. Its separate units will want to be different but, to survive, will need each other, will need to be part of a bigger whole so that they can command bigger resources.

The federal organization already exists in some multinational corporations. Their products are manufactured in many countries and distributed in many more by separate national companies, often with a majority of local shareholders. The headquarters of Philips may be in the Netherlands, of Nestlé in Switzerland, but neither would claim to be national companies of those countries. Their head offices have to juggle with local policy and overall objectives, seeking the best possible compromise. Local independence is demanded and granted but has to make necessary sacrifices for the good of the greater whole. Business conglomerates can also be federalist in nature, with a holding company keeping watch on long-term policy and finance while the individual units do their separate things.

Federalism fits well with the pressure for organizations to be more personal and human and therefore on a more human scale (see boxes 4.8, 4.9), which, loosely interpreted, means that their power and responsibility ought always to be pushed down as far as they can be in any organization.

BOX 4.8   SMALL IS PERSONAL

A meeting of the heads of nine successful British companies sought to identify any common reasons for their continuing success.

Common to all of them was the tradition of the four layers: in every plant, office or store there should, they all felt, be no more than four layers in the hierarchy, including the person at the top and the persons at the bottom. That meant no more than two intermediate levels of management and that in turn meant, in practical terms, a maximum of about 150 people per site. Growth, therefore, meant more sites not expansion of the present ones.

The second common aspiration was the desire to give everyone in the organization a feeling of ownership of the work. In several cases this meant, literally, legal or financial ownership of the firm; in other places it meant a sense of responsibility for the work produced, 'as if everything that went out had people's names on it'. 'If we can keep it small, we can keep it personal.'

Federalism therefore is, on the face of it, a good thing. It works too. The federal states of the world, be they large (the USA, the USSR, Canada, Australia and West Germany) or small (Switzerland, Yugoslavia) are among the most successful, while multinationals and conglomerates are so successful that, on occasion, they seem to threaten the power of sovereign states. Why, then, has federalism been so long in coming to ordinary organizations?

BOX 4.9   THE PRINCIPLE OF SUBSIDIARITY, AS RESTATED BY POPE PIUS XI

> Just as it is wrong to withdraw from the individual and commit to the community at large what private enterprise and industry can accomplish, so, too, it is an injustice, a grave evil and a disturbance of right order for a large and higher organization to arrogate to itself functions which can be performed efficiently by smaller and lower bodies . . .
>
> *Quadragesimo Anno*, 1941

There are two main reasons. First, unity and conformity have been the central thrust of most organizations, in business as in health, education or local government. That way the centre keeps control, by prescribing what has to be done and how, at the expense of diversity and independence; the fear has been that too much decentralization would lead to anarchy. Second, federations historically grow from the bottom up, as the separate groups come together for mutual advantage and protection. There have been few, if any, successful examples of top-down federalism, where a unitary system has turned into a federal system. Nobody willingly gives us power, and federal centres are less powerful than unitary ones. Who knows the name of the Swiss chief of state or the address of the head office of the International Postal Federation?

Federalism, however, becomes easier as communications improve. No longer do you have to prescribe in order to know what is happening; electronics allow you to know it in real time. One consistent finding in organization studies has been that more decentralization always needs more information. This is now easy. One block to federalism has been removed.

That being so, federalism becomes preferable to disintegration when the pressures for diversity grow too great. 'Things fall apart,' said Yeats, 'the centre cannot hold', but he was talking of a unitary state. A federal centre can encourage diversity in pursuit of common goals, and by limiting its control to a few essentials it can govern without oppression. We may therefore expect to see federalism become increasingly popular as an organizational form.

One outcome of more federal organizations will be a wider geographical spread of work, even if ownership becomes more concentrated, as many predict. This shift in work had already started in the 1970s (see box 4.10) and is thought still to be increasing. Unfortunately, the British have little experience of running federal organizations, having tended to regard federalism as a colonial invention. Indeed, the nature of federal constitutions runs

BOX 4.10 WHO NEEDS TOWNS?

Manufacturing employment by type of area 1960–78

	Employment (000s)		% change 1960–78
	1960	1978	
London	1,338	769	−42.5
Conurbations	2,282	1,677	−26.5
Free-standing cities	1,331	1,148	−13.8
Large towns	921	901	− 2.2
Small towns	1,631	1,887	+15.7
Rural areas	527	728	+38.0
GB	8,031	7,110	−11.5

Conurbations = Manchester, Merseyside, Clydeside, Tyneside, West Yorkshire, West Midlands
Free-standing cities = other cities with more than 250,000 people
Large towns = towns or cities with between 100,000 and 250,000 people
Small towns = local authority districts including at least one settlement with between 35,000 and 100,000 people
Rural areas = local authority districts in which all settlements have less than 35,000 people

Source: Department of Employment

counter to the British tradition in three ways. First, there is a formal written constitution, watched over by a trustee body (whereas the British prefer to rely on a changing body of case law); second, there are composite coalitions, representing the diversity of the nation states and horseshoe-shaped chambers to express the range of opinion (whereas the British prefer an adversarial tradition); third, the executive is separate from the legislature and therefore more governed by it (whereas the British prefer an executive which is the dominant part of the legislature).

The preference of the British for informal, adversarial and oligarchical government is expressed in their organizations as well as in their government. They will, no doubt, try to put a British gloss on federalism as it emerges, but they would be wiser to learn to work within the new tradition, for federalism seems to have its own imperatives. These could well emerge in the form of a revised Companies Act, imposing formal constitutions and participation on companies, in boards representing a wide array of interest groups, in executive boards with fixed-term contracts subordinate to supervisory boards (as in Germany and Sweden) – arrangements which will start in the business world and spread to other sectors as federalism becomes a more common form of organization.

The professional organization

The repetitive bits of work will be progressively automated. The accessory bits to the core of the organization's work will be gradually contracted out. What will be left? The specialists, the experts and the co-ordinators with a few helpers and dial watchers. More and more the organizations of industry and business will come to resemble organizations of professionals – schools and universities, hospitals and research departments – organizations built around individuals doing their professional thing. That sounds fine for the professionals, but what of those who are not professionals? They, it seems, will end up as the few helpers and dial watchers. The core of the large organization will not be big, and it will not be full of 'labour'. The days of the large employment organization are over.

BOX 4.11   PROFESSIONAL CREDITS

The notepaper of a professional firm – in law, accountancy, stockbroking or architecture – lists the names of all the partners. So today do the para-professions of advertising, management consulting and recruitment agencies list their directors.

The credits at the end of a television programme or a film are another outward and visible sign of a professional organization. Every individual who has contributed his or her special skills gets a mention and a listing.

Will the day come when our cars come signed by the workers who made them, as prints are signed and numbered by the artist?

In the large organizations of tomorrow everyone will be a professional of sorts, a 'something' as well as a 'somebody', just as one is a teacher, a doctor or a lawyer as well as an employee of a school or hospital. To say that a university is the model of the organization of tomorrow would be to exaggerate as well as to alarm us unnecessarily, but assuredly the organizations of business, and of government too, will have to come to terms with the problems of running organizations of professionals, among them the following:

(1) professionals demand a lot of independence and autonomy, the right to do their specialist task in their own way;

(2) professional organizations are flat structures, with perhaps three steps in the formal hierarchy;

(3) a professional career means advancement in one's profession, not necessarily in the same organization – professionals are mobile;

(4) on the other hand, professionals have effective tenure. It is hard for organizations to get rid of them except on the ground of unprofessional conduct;

(5) professionals prefer networks rather than formal bureaucracy;

(6) professionals prefer to be managed by fellow professionals but regard management and administration as a chore;

(7) professionals train the next generation of themselves. Professional organizations have therefore to be schools as well as work organizations.

In general the balance of power in professional organizations lies with the individual, particularly with the group of senior professionals. We may expect to see the organizations of business trying to alter that balance as they grapple with the realities of the professional organization, in particular:

(1) fixed-term contracts will become more common, as firms refuse to take on the responsibility of employing an expensive specialist for forty years or more;

(2) flexible time contracts will be introduced to make part-time work more feasible when an organization does not need the full-time services of certain specialists, many of whom will be pleased to be self-employed for part of their time;

(3) personal and portable pension schemes will become more common in order to facilitate the moves above;

(4) education and training will become an ever-increasing priority in organizations, partly because they will be best placed to do it, partly because training will be a way of harmonizing the interests of the individual being trained with the interests of the organization.

In spite, however, of moves to make the terms of employment more flexible, the organization of the future will consist of a professional core and a high degree of automation. Some of this automation may well be unwelcome to the new professionals because it will force them to learn new skills and will deprive them of their customary support staff (see box 4.12), but productivity has to embrace professionals as well as more ordinary mortals.

BOX 4.12   AN EXECUTIVE JOB SHOP

A professional manager was boasting of his electronic mail system. 'Come round and see it,' he said. When I got to his office there were no secretaries to be seen anywhere. Instead where his secretary used to sit there was a keyboard and a VDU. He pressed a key and up on the screen appeared a series of messages — what would in the past have been memos — from different parts of the organization. Instant communication! He beamed with pride and then pulled out a chair and sat down in front of the keyboard.

'What are you doing?' I asked.

'Well, now I have to reply to them,' he said, laboriously and slowly starting to type his replies with one finger.

The new executive has to be good at many things when there is no one to whom to delegate. Deskilling is the painful start to new learning (and better productivity?).

Professional organizations have to be run by consent rather than command. You cannot tell people what to do; you can only 'ask' them. A greater proportion of professionals in the organization will increase the pressures for participation and democracy in policy-making. Nor need any manager think that the automation of routine work and the subcontracting of ancillary work will get rid of unions, for the unions of professionals are traditionally more intransigent and more monopolistic than their craft equivalents, even though they may call themselves associations rather than unions. We may expect to see a move by the newer professional associations to become more like guilds than professional societies in order to underline the protective purpose of the body.

But the greatest problem associated with the growth of the professional organization will be its accentuation of the gap between the professional and the ordinary workers left in the organization, the people who will still be needed to watch and feed the equipment, wheel some of the trolleys, supervise school meals, clean and tend to patients, man the reception desks, file the letters and answer the telephones. Try as they may, organizations, and particularly service organizations, are not going to be able to do without human beings to carry out these mundane but important tasks, certainly for many years to come. The professional organization is likely to have two sorts of people – the knowledge people (the professionals) and the muscle people (the 'hands', the ancillary staff). Missing will be the people in between – the skilled worker, the foremen, the middle manager. If we are not careful, organizations will become two-class societies even more than they are already. The divide is aptly summed up by the title of Mike Cooley's book *Architect or Bee*,[2] in which he points out forcibly how deskilling new technology can be, in spite of all its potential for good, if it is regarded as a machine serviced by man rather than as a tool in the service of man.

If the professional organization does not become a single-status organization, in which everyone has the same rights and benefits, we may expect to see the unions representing the ancillary staff adopt a much more aggressive stance because even in a weak labour market these groups of people command considerable negative power. They can hold the organization to ransom quite effectively and will use that power, if driven to it, to bid up the commodity price of their work.

What can the organization give these essential people other than money? How can it prevent their becoming the wage slaves of the new society? Only, I would suggest, by giving them some kind of stake in the organization. This could be a financial stake, through share ownership or profit-sharing, or it could be an emotional stake by building them into small teams working with the professionals, integrated groups rather than a layered and functionalized organization.

The Americans have a phrase, 'hi-tech needs hi-touch', meaning that the dehumanizing element of much of the new do-it-for-you technology has to be compensated for by accentuating other ways of relating to people.[3] People want to be with people. If we are locked into our computer terminals all day, in the home or the office, we shall socialize more intensively at night. We shall seek automated shopping for some things, perhaps, but more personal service for others. Small restaurants will flourish alongside the fast-food batteries. So in factories the robots must be accompanied by the currently fashionable 'quality circles', teams in job shops and pressures for smaller factories and offices. Our increasingly professional organizations would be wise to remember that people need people in inverse proportion to their ability, technically, to do without them. The professional organization must also be the small group organization, the participating organization, the democratic organization.

'Japanese drift?'

These consequences add up to a trend which might be called 'Japanese drift'. Slowly but surely our organizations are beginning to look more Japanese. That does not mean that everyone is doing physical jerks together at the beginning of the day or singing the company song; what it does mean is that the way things are organized is beginning to look more and more like the way things are organized in Japan.

There is a myth that everyone has a guarantee to lifetime employment in Japan. Not so. To begin with, any such guarantee is technically illegal. The Japanese Labour Standard Law states that 'Labour contracts cannot be held for a period extending more than one year.' Any guarantee is therefore informal. Furthermore, the law applies only to 30 per cent of the Japanese workforce, those working for the large integrated corporations.[4] Finally, guaranteed employment generally ends at 55, five years before the state retirement age.

The truth is that the large Japanese organizations float on a raft of small subcontractors (see box 4.13). Half of the entire manufacturing output in Japan is made by small subcontractors for large enterprises, that is, companies with more than 300 employees.[5] Japan is a country of small businesses which use low-cost labour and do not offer life-time employment. This raft of small businesses allows the large enterprises to concentrate on design, marketing and final assembly. The small businesses absorb the fluctuations in demand and in the need for labour as their big brothers vary their orders, but in return they do not have to bother about marketing, since something like 70 per cent of their output goes to the larger enterprises. These small firms also pick up the employees of the larger firms when they reach the age of 55.

Japan also has a small army of the self-employed (see box 4.13), including a group euphemistically called 'unpaid domestic workers', the unpaid members of the family who help in the one-man business. Any reduction in hours worked by these self-employed, or by the workers in the subcontracting firms, is hardly visible to the outside

BOX 4.13　SUBCONTRACTING AND SELF-EMPLOYMENT IN JAPAN

'We found many large enterprises on top of a huge triangle of small enterprises. The Japanese car maker provides a typical example: on average, a Japanese car maker has 36,000 sub-contractors, of which all but a thousand are small enterprises.'[6]

BOX 4.13 continued

Subcontracting structure in the automobile industry

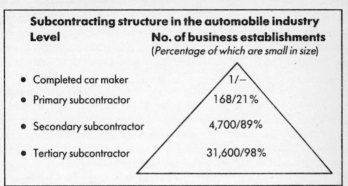

Subcontracting structure in the automobile industry	
Level	**No. of business establishments** *(Percentage of which are small in size)*
• Completed car maker	1/—
• Primary subcontractor	168/21%
• Secondary subcontractor	4,700/89%
• Tertiary subcontractor	31,600/98%

Source: SMEA, *Survey on the Structure of Specialization (Automobile)*, 1977.

The composition of the work force of small enterprises enables them to absorb the excess labor problems of the large enterprises to a certain extent. The large number of self-employed workers can reduce their own workload or that of unpaid family employees.

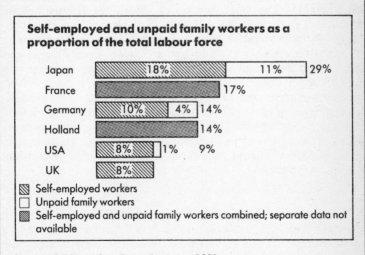

Self-employed and unpaid family workers as a proportion of the total labour force

Japan	18% / 11%	29%
France	17%	
Germany	10% / 4%	14%
Holland	14%	
USA	8% / 1%	9%
UK	8%	

▨ Self-employed workers
☐ Unpaid family workers
▦ Self-employed and unpaid family workers combined; separate data not available

Source: OECD, Labor Force Statistics, 1981.

world. Of those working in these small businesses 8 per cent have no hours specified in their contracts.[7] *Under*employment is not the same as *un*employment.

How did this happen in Japan? After the Second World War the American occupation forces broke up the large organizations into smaller units. These units lacked the capital to do everything themselves and so contracted out bits of the work to smaller units. Low interest rates and a land-redistribution policy allowed these smaller businesses to raise the smaller amounts of money they needed in order to finance themselves, once they had the guarantee of a contract from one of the 'big boys'.

The cultural side of things is also important. Although competition among subcontractors is encouraged, it is not relentless. There is a strong bond between subcontractor and large enterprise – it is in both their interests that the subcontractor should be efficient, competitive and independent (as Marks and Spencer has demonstrated in this country).

The smaller businesses are relatively untouched by the unions, because the unions in Japan are company unions, organized around the larger employers. This allows the smaller businesses to establish a competitive edge by working longer hours, if their workers agree. This encourages the larger organizations to contract out their labour-intensive work to the smaller firms. The labour law allows an enterprise to operate forty-eight hours a week before paying overtime, although its public objective is to lower this figure to forty hours a week.

Will it happen in Britain? If you put together the contractual dispersed organization and the increasingly professional nature of the larger enterprise, it could be argued that it is already happening. The large organizations are cutting back their staff to an essential core to whom they are prepared to offer an effective, if not a specific, guarantee of continued employment. The terms and conditions of this core staff will be tightly negotiated and enforced by the unions in a last flexing of their bargaining muscle. To compensate, the organization will continue to subcontract as much as possible of its labour-intensive work to small dependent businesses and self-employed skilled people, who will be able and willing to work longer for lower costs because more of the fruits of success go into their own pockets.

As self-employment grows, so will the unpaid domestic activity (the black and grey economies) that supports this part of the official white economy. Underemployment, in the self-employment economy, will be as unrecorded in Britain as in Japan. Part-time working is and will increase.

Badly handled, this 'Japanese drift' is a recipe for exploitation. The large organizations could export all their uncertainty, all their fluctuations in the market or their order book, to their small subcontractors. They could use competition ruthlessly to exploit any short-term advantages of one subcontractor over another, killing the capacity to invest or grow in the longer term. The new self-employed could turn into the technological out-workers of the twenty-first century, exploited just as much as the knitters and seamstresses of yesterday. High interest rates and the lack of any longer-risk capital could starve the smaller firm of investment, forcing it into an unhealthy dependence on short-term cash flow, doing anything to keep the cash coming in. Some of this kind of corporate exploitation undoubtedly happens in Japan. In times of recession work is pulled back into the big corporations, stranding the smaller and the weaker. When money is tight the working stock is carried, and paid for, by the subcontractor, who is asked to deliver precise amounts to the factory gate at precise times, even to the extent of a delivery every four hours. Retirement at 55 is possible for many only because subcontractors are obliged to find them a home for five years if they want to keep the business.

There is a very real danger that Britain may drift into this two-tiered structure. It is too easy to export the organizational uncertainty to other, less well protected shoulders. It is an obvious route to tighter economics, even if it does require more complex management skills. How, then, do we stop or at least control the drift?

One answer may be through the spread of co-operatives. The extended family business in Japan is unlikely to find its counterpart in Britain today, where the family is a domestic rather than a work unit. The co-operative, whether achieved by buy-out (see box 4.14) or by partnership arrangement, is a way of spreading the overheads and sharing the risk among a *work* family. Britain's record in the co-operative field has not been illustrious and is known more for the failures, like Meriden, than for the successes. But that may be because Britain's co-operatives have tended to be ideologically rooted in socialism to such a degree that they have turned their backs on the realities of capitalism and the market society in which they are forced to operate. The less dogmatic co-operatives of the Italian and French construction industries are, as Jenny Thornley demonstrates,[8] very successful in blending commercial success with common ownership and democratic practices. The new co-operatives of the subcontractors may be forged more out of business necessity than out of ideology, and this could be a better base for their success. They will, however, be wise if they learn to live in the shadow and under the patronage of the large corporations, for such patronage will ease their marketing problems, help with their planning and even underwrite their finances.

For its part, the large corporation has to act responsibly rather than narrowly and selfishly. Subcontracting, like delegation, demands responsibility from both sides. Large corporations should be encouraged to think of themselves as patrons, providing an umbrella of support for the fledgling groups working for them. Laws and regulations may help. Government support and guarantees would not come amiss, but the best trustees of the future of the growing world of small subcontracting business will be the large corporations. Responsibility cannot be imposed on them by law; it has to be voluntary. Survival in a tough economic world will be the driving force behind the moves to more dispersed, contractual and federal organizations, but survival will not lead to future growth unless it is responsibility for the survival of others.

BOX 4.14　MANAGEMENT SPIN-OFFS

Another form of dispersal is the spin-off, the buy-out or the demerger. Because this expedient makes financial sense, or because it is the alternative to bankruptcy or redundancy, or because it is part of a deliberate policy of development, large organizations are increasingly eager or willing to hive off part of their operations to be run and owned independently. Recent examples include:

Panache Upholstery – in 1980 Bob Wilkins, its managing director, persuaded sixty employees to invest their redundancy money to rescue this offshoot of A & M Upholstery, which would otherwise have closed it.

National Freight Corporation – in 1982 the employees of this nationalized courier raised £53 million to buy it from the government in accord with its privatization policy.

Options – a young editor of IPC set up a firm of journalists to produce magazines to be promoted and distributed by IPC; *Options*, the women's magazine is one result.

Control Data Corporation has a policy and a special fund to help employees who set up on their own, although by mid-1982 only one firm, in computer training, had been created in Britain.

Of course, 'breakaways' have long been part of the tradition of advertising, consultancy and finance – not always, however, with the approval and involvement of the parent company.

'Japanese drift' could easily be a recipe for an economic jungle, with the large corporation as the lion, or it could be the impetus behind a move to smaller, more open and more democratic units of work, where people are more their own masters, where labour can be combined with dignity and work seems less of a chore. Whether it goes that way or not will depend very largely on whether those who lead the way see themselves as clearing the undergrowth or planting new trees.

5

The New Agenda

We stand on a hinge of time. The door is closing on part of our past and opening on to a new future. Dramatic stuff, but it does not always feel that dramatic when it is happening, for hinges can be painful places if you are caught in them. Moreover, hinges do not themselves have to move a lot to open up quite a new scene. Those, therefore, who stand on hinges are not always best positioned to glimpse the promised land or to chart the way ahead. Too easily they are mesmerized by the problems and pain of the move and, if anything, would like to shut the door on the future and keep hold of the past that they knew.

Maybe that is why there is so little political debate about the implications of the new scenario that has been sketched out in the previous chapters. All the argument is about how to stop the change, reverse it or at least slow it down. Of course, politicians are psychologically tuned into the short term. Three years is just about as far ahead as a Government likes to think, knowing (or believing) that if the payoff is not showing through by then, it will lose the support of the voters. The agenda thrown up by the new scenario for work is not going to produce much voter satisfaction within the next three years – indeed, it is going to cost most people something. Perhaps it is not surprising, therefore, that politicians shy away from the agenda, preferring to hope that the need for it will go away or will at least be postponed until after their term of office.

The result has been a strange conspiracy of silence, revealed most obviously by the 1983 British general election. The man, woman or young person in the street knows that the world of work has changed, probably for ever. Few people really believe that full employment will return, that the forty-seven-year job will be the norm again or that high rates of economic growth will again allow us to have low taxes and increased public services. Yet the election was fought on the basis that all these things were possible if only you voted for the right party. Pensions, education, state benefits, taxation policy – those parts of public policy which are crucial to any discussion of the future of work – were barely mentioned.

For a decade now unemployment has been above the 3 per cent figure that Beveridge suggested was appropriate for effective full employment. For twenty years manufacturing jobs have been declining. Yet neither Labour nor Conservative Governments have done more than offer short-term palliatives. If a future Government were courageous enough to face up to the long-term changes in the nature of work in our society, it might well find that it was moving with the stream. People do need answers to the new questions, even if they may not always like the answers they get. But then perhaps the main lesson of the 1983 election was that the British voter is not always put off by the unpalatable if he or she believes it to be necessary. It demeans the average citizen to regard him or her as always a short-term maximizer. If history proves anything, it is the opposite – that the average man or woman thinks of children and grandchildren, while politicians think of the next election. Society can stand more truth than it is offered. It is time to put the new questions on the political agenda so that all concerned – employers, unions, teachers and parents as well as governments – can begin to look for answers.

In chapter 1 we glimpsed part of the new agenda. In this chapter we can flesh out some of the more important of the new questions. To answer each of them fully, or even to explore all of the options, would require a separate book on each. Some of those books are already beginning to be written. This chapter will help to put them in perspective, for the answers to the different questions cannot be considered in isolation; they all form part of a changing pattern, in which the answer to one affects the answer to the next.

There are four big new questions:

(1) Who gets the jobs? Jobs may be fewer, shorter and different, but some way of allocating them is still needed. What counts as a job?

(2) How do we pay ourselves? If we do not all have a job for life, how do we get our income? Who will pay for whom, and how will money be collected and distributed?

(3) What do we use for wealth? How much wealth do we need as a nation and how do we get it? What needs to be done, by whom, to make more of it? What counts as wealth anyway?

(4) Who protects us? Consumers and workers will both need new forms of protection in the new, more dispersed world of work. How is this to be organized? What will be the role of the unions?

Big though these questions are, there are two that are bigger still: How do we prepare people for this new scenario? How do we educate people for tomorrow? These questions need a chapter to themselves.

BOX 5.1   WHAT DO WE MEAN BY FULL EMPLOYMENT?

The fact that 5-year-old boys, filthy and frightened, can no longer find employment climbing chimneys and 8-year old girls can no longer crawl through the darkness and dust of our pits dragging great cartloads of coal like wretched animals does not (by any sane definition) mean an appalling rise in child unemployment which must be reversed at once if not sooner.

The fact that hundreds of thousands of teenagers are staying on into the sixth form or attending polys and universities is not some cynical trick to hide the true level of unemployment.

The fact that the basic working week today is forty hours or slightly less, compared with a basic eighty hours, is not an indication that we are suffering, Third World-style, from 'concealed unemployment' because two people are nowadays doing a job which ought to be performed by one.

John Torode
Guardian, 1 September 1983

WHO GETS THE JOBS?

Even if, as chapter 3 argued, jobs get smaller, there will not be that many more of them, at least not in the conventional organization. If jobs do not get smaller, there will be even fewer of them. It is therefore important that jobs shrink to 50,000 hours without loss of productivity and that they are fairly and efficiently shared out. That will not happen automatically. There are too many small monopolies in the labour market, too many people quite legitimately concerned to look after themselves and to let the country, in some way, look after the rest.

The key issue here is going to be the length of a job life. If a thirty-five-year job replaces a forty-seven-year job, then twelve years have to be lost somewhere – probably at the beginning or the end. But who is going willingly to give up twelve years of a job unless something is offered in exchange? If things are left to themselves, those who have jobs will cling on to them, and the twelve years will have to come off the beginning of the job. Entry to the labour force will be effectively postponed. That may be fine if you can fill the intervening years with higher education, professional studies and the occasional interlude in Nepal and India with guitar and friends, but not all are so fortunate, and ten years in the back streets of our cities with the occasional interlude on a Manpower Services Commission course is hardly a satisfactory alternative to a job. In 1983, 300,000 young people under 24 had had no job for more than a year. It is happening now.

If the shrinking job is to be a positive benefit to all, positive alternatives must be offered both at the beginning and at the end. The missing years cannot, with any justice, be left to the jungle of the labour market to share out. There is a range of options.

Extended youth education

A commitment to educate all young people to age 21, and not just those who were clever enough to go to college, would cut the average working life by five years. The education would have to be 'appropriate' to the interests and probable occupations of the young people and would therefore, in most cases, mean some version of work-related training rather than the knowledge-based programmes of universities.

The British Youth Training Scheme is one step in this direction, with its attempt to link work experience and training, but it would need to be extended to four years and to set itself higher goals in terms of skills and qualifications if it were to make a significant impact. In its early stages there are bound to be many teething problems – too many inadequate courses, too many skills learned just in time for them to be redundant in a changing technology, too many bad experiences of work. We must hope that there are sufficient good experiences to allow the scheme to be extended.

It is some sort of national disgrace that 44 per cent of Britain's young people go into a job today with no qualification at all, compared with 6 per cent in West Germany. In a world of work where a saleable skill is going to be even more essential, where mere bodies and muscles are no longer enough, the lack of any qualification is tantamount to a life sentence of insecurity. The thirteen weeks of training guaranteed by the Youth Training Scheme is like a thimble of water to a man dying of thirst – only to be justified if it is intended as a foretaste of much more to come. But much more would, of course, cost much more – perhaps £4 billion p.a., compared with the £1 billion of the Youth Training Scheme in its first year.

Easing exit from employment

Most people would be willing, even happy, to leave employment after thirty-five years if they could afford to. The TUC would like to see the male retirement age brought down to 60 (as the French intend to do, and as the Italians have already done), with a minimum pension level equivalent to half the gross earnings of a married couple. In 1980 it was calculated that this would cost an extra £9 billion. Although the TUC proposals are not out of line with the provisions that already exist in some other European countries, they are too expensive to be implemented across the board. What we can expect to see are more pressures for:

(1)   a general increase in the minimum level of pensions, particularly for those (in their fifties and sixties) who will not benefit greatly from the 1975 Act providing for the start of a national pension scheme related to final earnings;

(2)   equalization of men's retirement age (65) and that of women, probably by meeting in the middle;

(3)   better provision for flexible retirement, with adjustable pensions.

It is this last option which provides most scope for imaginative action and enabling legislation.

Why could it not be made easier for the older workers to work part-time? Pension rules could be adapted (see box 5.2) to allow for more part-pensions. Firms and voluntary organizations could join forces to provide two part-time jobs for older workers, one paid, the other covering expenses only, with full pension guaranteed at final retirement. After all, if a pension is protected, many people in their late fifties need less money than they did ten years earlier.

Why cannot pensons be made more transferable? As occupational pension schemes are currently organized, we lose out every time we change jobs. The employers we leave would have to put something like the whole of an extra year's salary into our pension schemes to preserve the final value of our pensions. Few are altruistic enough, and few of us have the bargaining power to force a new employer to do it instead. A law requiring pension schemes to be topped up when an employee leaves would, in effect, penalize the employees who remain and would therefore seem inequitable. Personal and portable schemes, which individuals take out with an insurance company of their choice and which are paid for by employers would, on past evidence, produce a lower pension than the two-thirds of final salary guaranteed to those who worked forty years in one firm.

All the pressures are to stay put and stay long. It will, however, be increasingly in the interest of firms and individuals to make pensions more personal and more portable in order to give both sides more freedom, choice and flexibility.

BOX 5.2   PART-PENSIONS

In 1976 the standard retirement age in Sweden was 65. Employees between 60 and 64 who reduced their working time by at least five hours but continued to work at least seventeen hours received a pension equivalent to two-thirds of the earnings they gave up, in addition to their part-time salary. Over 200,000 workers were eligible in 1976, and 53,000 were in the scheme.

A similar scheme has been proposed in Britain (by Michael Pilch and Ben Carroll in 1977).[1] The ideal is to create a flexible retirement bracket for both men and women between 60 and 70. People could choose half-pensions between 60 and 65 or a tax-free cash bonus of 30 per cent of their salaries. The authors estimated that it might cost an extra £1 billion in full operation.

More fixed-term contracts

The implication of 'tenure', or a job for life as long as the business lasts, is built into current employment legislation. It does not apply to fee-based contracts. We shall see more of these contracts, as chapter 4 argued, and this will in itself provide more flexibility in determining the length of working lives. If more contracts of employment could be based on a fixed number of years, rather as happens in the armed services, and pension arrangements adjusted accordingly, it would free the working lives of many people, mainly the ones with professional or managerial qualifications who are best able to work as self-employed, outside the job market.

Cheaper youth wages

There is some evidence that youth labour has been overpriced in the past (see box 5.3), keeping young people out of the workforce in preference to older, more experienced workers who are not that much more expensive. The British 'Young Workers Scheme' is designed to change this by paying employers a subsidy for keeping down youth wages. There are also fears that some employers will use the Youth Training Scheme as a way of obtaining cheap young labour.

Youth wages would have to be very low to add to the total numbers employed. Slightly lower wages mean only that youth is employed at the expense of older workers. If this happens without proper compensation to the older worker, it must be seen as a device by which nobody wins except the employer. It is more acceptable if it is linked to proper arrangements for earlier and flexible retirement policies and to proper training, followed by proper wages, for young people. In that way the unskilled, untrained young person could price himself or herself into work without prejudice to others.

BOX 5.3   IS YOUTH TOO EXPENSIVE?

A Department of Employment study (reported in the *Department of Employment Gazette* for June 1983) found that if the earnings of young people increased by 1 per cent more than adult earnings, then youth employment increased by 2 per cent over and above the effect of the recession.

In the 1970s youth earnings rose from 47 per cent of adult earnings to 56 per cent, although they have since fallen back to 53 per cent. A quarter of all young people under 18 were without a job in the summer of 1983, double the proportion for the workforce as a whole.

The implication drawn by many is that lower youth wages would result in more jobs for young people – but presumably fewer jobs for their elders.

The danger of all this lies in the possibility that the traditional job market will remain a jungle. Those who have the jobs will cling on to them, helped and encouraged by their unions. Because their organizations will not be growing in terms of people, that will leave fewer opportunities for those in the middle to advance and no opportunity for those on the outside to get in until they are in their mid-twenties. Do we really want to see our mainstream organizations become increasingly middle-aged and full of yesterday's skills? Do we really want to see a raft of young people waiting around until their early twenties with nothing to do apart from one year on the Youth Training Scheme?

It is not just a matter of occupying the young. If the core organizations become dull, middle-aged and ossified, the talent will go elsewhere. It will go to the new areas of work (see box 5.4), many of which will be young, small, free and experimental, the exciting fringe of society. Do we, however, want a society with an exotic fringe and a moribund core? Can such a society survive?

BOX 5.4   THE NEW JOBS

Barry Jones, in Australia, identified sixteen areas for 'new work', not all of them, as he observes, desirable:

(1) education, including recurrent education and training for the semi-skilled and unskilled;

(2) home-based employment, including domestic work, maintenance and gardening on a contract basis, home security;

(3) leisure, tourism, sport and gambling;

(4) dining out;

(5) provision of drink, drugs, and commercial sex (and treating their adverse effects);

(6) craftwork, the arts and entertainment generally;

(7) individualized transport systems: taxis, personal drivers, fixed route minibuses, courier services, point-to-point delivery;

(8) public-sector employment: administration, armed forces, the police;

(9) hobby-related work, including DIY work in the informal economy, antiques and collecting;

(10) individualized social, welfare and counselling services (especially geriatric or psychiatric);

(11) small-unit energy generation (solar, wind, and growing crops for 'biomass') and subsistence farming;

(12) manufacture of leisure and solar-energy equipment (boats, games, solar heaters and collectors);

(13) materials recycling;

(14) 'welfare industries', where the main output is employment;

(15) nature-related work: gardening, care and preservation of wilderness, forests, deserts and natural parks, coastlines, footpaths;

(16) care of animals, including selling, breeding and grooming pets.

Barry Jones
Sleepers Wake! (Wheatsheaf Books, Brighton, 1982)

Will this new work suck up our new talent? Should it?

The traditional employment organization may play a smaller part in the world of tomorrow, but it will still be crucial, whether it be the business, the department store, the school or the government office. Jobs will always matter. We must make sure that they are attractive to the young and the competent, for whom we must make room. We must train them. We must reward them properly. Left to itself, it is unlikely that the job market will respond in the right long-term way. The question has to get on to the political agenda if our organizations are not to become the waste bins of the new society.

HOW DO WE PAY OURSELVES?

Full employment solves many problems, not least the question of how to distribute wealth, cash, so that everyone has the means of access to the goods and services of society. The guarantee of a job is also the guarantee of a wage. There may be argument about the size of the wage, but at least the basic mechanism is there. Under full employment the role of the welfare state is to provide a safety net for the unfortunate and short-term measures to support those temporarily without a job. Because joblessness is temporary, society can afford to be either generous (as in many European countries, providing benefits almost equivalent to a full wage) or mean (confining itself to subsistence levels of support) because either way it will not last long.

The whole equation changes if full employment can no longer be assumed. The job is no longer an adequate or fair mechanism for the distribution of wealth, partly because some will have no job for long periods, partly because there will be a great disparity between the different types of job. Those backed up by automation and capital expenditure could justify high wages; those in the new personal services may be less fortunate.

Karl Marx had something to say about such a state of affairs, although he thought it would happen a hundred years earlier. Marx believed that 'technology necessarily displaces workers in the branches of industry into which it is introduced' but accepted that the same process might also 'lead to an increase in employment in other branches of industry'. Gradually, however, skilled workers would be replaced by machines, and the only employment increase would be unproductive work, particularly domestic service. Marx predicted the growth of a semi-employed category of workers, receiving low wages for occasional work, and an increase in the number of people dependent on the Poor Law. He argued that women and children would find more employment and skilled men less as automation advanced. As Bill Jordan points out,[2] Marx modified his views at the time when the facts did not start to follow his theories, but he might have found the contemporary scene understandable. Marx's fear was that increased productivity would create not a more leisured society but a more unequal one.

That has to be our concern today. The availability of jobs for some and work for all (of different kinds) is no guarantee of wealth for all. Society has to find a difficult balance between rewarding the wealth creators and distributing wealth fairly to all its citizens. If it does not, there will increasingly be two societies – those in well paid jobs and those skirmishing, more or less successfully, on the outer fringes. The question of how we pay ourselves to live in a society where, to follow Professor Stonier,[3] only 10 per cent are needed to provide all our essential needs is perhaps the most crucial question of the future of work. There are, once again, a range of options.

Pensions earned during working years

A pension should, in theory, be deferred pay, saved and invested on employees' behalf until they retire. In practice it has been interpreted, by individuals, states and organizations, as an obligation to continue to provide a reduced wage or salary after retirement, in proportion to final earnings. That is equivalent to making the pension a charge upon one's successors, for few pension schemes are sufficiently well funded to cope with inflation, increasing lifespans and earlier retirement dates. The pension has proved to be an impossible obligation to fulfil, particularly for those who come late to an organization or leave early. For the majority of people the promise of their occupational pension schemes have turned out to be the con of a lifetime if they have moved or retired early.

The first option therefore requires a return to the original concept of a pension as deferred pay and implies that the individual and organization together put away enough during thirty-five years of work to live on during the twenty-five years after employment. This might mean contributions amounting to 50 per cent of one's salary, double the size of most contributions now, if one wanted to maintain the principle of a pension equivalent to two-thirds of final salary. Most individuals and organizations would find this ruinously expensive – a further pressure on their labour costs.

Alternatively, the two-thirds principle could be abandoned on the grounds that retired people need less money and have more scope for supplementing their pension by marginal work or gift work in their homes.

Either way, the responsibility for providing for the future would revert to the individual, with minimum levels prescribed by law. It would be yet another encouragement for the idea of personal pension schemes. The fear, of course, is that many would not provide adequately for themselves and would still end up as an unfunded charge on the state. The option also assumes (a) that everyone can get a proper thirty-five-year job and (b) that the job is available earlier rather than later in his or her working life. Neither of these can be guaranteed. Personal pensions cannot therefore be enough in themselves.

A national income scheme

If the state cannot guarantee a job which provides a livelihood, then, say some, the state should guarantee livelihood directly. After all, in most modern states today the state provides, as a universal right, access to education and health care. Why not, then, access to the basic necessities of life such as food, clothing, shelter and heat, by means of a straight cash payment?

In its simplest and purest form a national income scheme would replace all the present complex social security benefits by paying a straight cash sum to every individual, whatever his or her age, marital status, means or incapacities. This money would then be progressively clawed back from those with income. It would be a way of pushing the money around in society so that everyone was guaranteed a minimum. Schemes such as this have been mooted for some time, first by Major C. H. Douglas in the 1920s, and have been called variously a social wage, negative income tax, social dividend or citizen's wage.

The more visionary advocates of this sort of scheme see it as reducing both the demand for formal jobs (because some second earners in families would no longer need or want a job) and the cost of labour, and therefore of prices on goods, because the national income element would be taken into account and would promote a genuinely freer labour market.[4] In the short term, however, taxes would have to be raised enormously to collect enough money to redistribute. VAT rates of 100 per cent have been mentioned and basic income tax rates of 50 per cent or more. One attempt to cost a scheme is given in box 5.5.

While a full national income scheme seems too big a step to take, the principle behind it is important – that a citizen is entitled to a livelihood as much as to education and health care, that the state is not a provider of last resort to those who cannot help themselves but a guarantor of basic rights.

BOX 5.5   COSTING A NATIONAL INCOME SCHEME

One approximate (and optimistic?) costing of a national income scheme was prepared by the Political Committee of the Ecology Party in their pamphlet 'Working for a Future', using 1980–1 data.

Basic payment rates (to individuals, irrespective of means or marital status)

Rate	Ages	Numbers involved (millions)	£ per week	Total cost (£m)
A	0–5	3.5	7.30	1,330
B	5–18	11.5	10.00	5,980
C	18–60	29.0	20.00	30,160
D	60+	12.0	28.00	17,470

Cost of basic payments	54,940
Cost of housing addition payments (to cover local authority rents)	7,500
Gross cost of scheme	62,440
Less All social security spending	25,000
Tax and mortgage allowances	13,000
Savings in government administration	4,500
Net cost of scheme	19,940
Cost to be covered by increased revenue income tax @ 40p	3,000
higher indirect taxation	7,000
Redirection of government spending	10,000
	20,000

Insulating benefits

The third option is a half-way stage to a national income scheme. It would work within the present benefit structure but would try to remove many of the disincentives to extra work which arise because of the abrupt way in which benefits disappear if income increases – the 'poverty trap'.

For instance, unemployment benefit disappears if one earns more than £2 per day. Supplementary benefit is not payable if one has savings of more than £2,000. These are hardly inducements to the unemployed to create their own jobs or to set up businesses with their own capital. There needs to be a higher ceiling above benefits, a guaranteed period for payment and a more gradual system for easing them out. The benefits need to be insulated, for a time, against success (see box 5.6).

BOX 5.6　GUY DAUNCEY'S PARADOX

In his 'Unemployment Handbook' Guy Dauncey looked at the problem of insulating benefits. He proposed:

The system to be changed so that claimants keep the freedom to earn money. This change could happen through a system whereby a person's *entitlement* to benefit is *increased* the *more* money you earn. This has a marvellous paradoxical logic to it, which makes sense as soon as you put figures to it. Benefit entitlement is increased by 50 per cent of the amount that a person earns. (That's the basic rule.) Thus a person on a basic benefit of £40 per week who earned £20 of her own accord while 'unemployed' would have her benefit entitlement increased by £10 to £50 per week. Having earned £20 herself, she would receive £30 in benefit. If she earned £40, her entitlement would increase to £60, and she would receive £20 in benefit. When she earned £80, she would become free of the benefit system altogether. In this way, people could work their way towards self-employment or towards the setting up of a small co-operative while unemployed, an impossibility under the present system.

G. Dauncey
'An Unemployment Handbook', National Extension College,
1981

Similarly, because marginal work is added to existing income, the marginal rates of taxation on any form of part-time self-employment are high, encouraging some people to trade in the black economy rather than to declare it and others to see the post-tax earnings as too small to justify the risk of a new venture, however small. It would cost the state almost nothing to treat part-time self-employment as a separate tax entity, with its own start-up or personal allowances, because the state collects almost nothing from such work as it is.

Here again, the philosophy is more important than the technical rules. Benefits should be viewed not as an alternative to income but as the guaranteed and insulated base of an income for a fixed period. Benefits would not be seen as the alternative to work, a scrounger's charter, if in the early stages income from work was an addition, not a replacement. When work is regarded as a replacement for benefit, then clearly people are going to calculate whether or not it pays them to replace benefit by work (their 'replacement ratio'), and some will calculate that work does not pay (see box 5.7).

Reducing benefits will fuel desperation and push more people to search harder for work. This is one way. It might work if there were jobs waiting to be filled. Without those jobs it is imposed poverty for

BOX 5.7 IS IT BETTER ON THE DOLE?

Professor Patrick Minford of Liverpool University suggests in his 1983 book *Unemployment – cause and cure* that 15 per cent of the workforce would be better off unemployed.

The Institute of Fiscal Studies, however, believes this figure to be greatly exaggerated because it overestimates the costs of going to work and underestimates the benefits available to low-paid people in jobs. They prefer a figure of 2 per cent, with perhaps another 5 per cent in a neutral position, neither worse nor better off.

A study of unemployed men, undertaken for the Inland Revenue, showed that 9 per cent had higher incomes than when at work but that for 35 per cent income when unemployed was less than half that in their last job. The Inland Revenue concluded that the evidence 'does not suggest a close relationship between high replacement ratios and job search behaviour' (reported in *New Society* 16 June 1983).

The United Nations Economic Commission for Europe reckoned (in 1983) that the disposable income of a married worker with two children fell by 47 per cent when he became unemployed in Britain, compared with only 8 per cent in Denmark and 10 per cent in France, Holland and Ireland.

Researchers at the London School of Economics, in a discussion paper of the Centre for Labour Economics, suggest that a 10 per cent drop in unemployment benefit might push 90,000 back to work. Professor Minford thinks that the figure would be 700,000 but does not suggest what jobs they would find.

In 1983 721,000 young people on the dole were living with their parents. If they were over 18, they received £23.65 a week. If they got a job away from home, they needed to replace this income and earn enough for their housing. They then had a high 'replacement ratio' or little incentive to take a low-paid job. On the other hand, if they continued to live at home, the 'replacement ratio' was low – even a poorly-paid job made them much richer. Where you live is critical.

many. Insulating benefits is another way, a way which would encourage more people to search for income opportunities through self-employment, making work, not looking for work. It is a more positive, more thrusting, more creative and ultimately more revenue-producing way to support one's fellow citizens when out of work and wanting to create work.

The Manpower Services Commission has invented its own insulating mechanism – a salary for any unemployed person who has £1,000 and a good business idea while he or she gets it started. It is an ingenious way of paying benefit to a start-up entrepreneur. The principle is excellent, but why need it be so complicated and tied up in such bureaucracy?

BOX 5.8   COULD WE CHANGE OUR TAXES?

A subcommittee of the House of Commons Treasury and Civil Service Committee examined possible ways of reforming Britain's tax and benefit system to reduce the 'poverty trap' and the burden of taxation on the poorer members of the population. They considered several possible packages.

(1)   The 'high-threshold' package which raised the tax threshold and relieved many of all income taxes. To raise the threshold by 28 per cent would mean that the basic rate of tax would have to go up to 46 per cent, which would in turn make the 'poverty trap' more severe for those on the edge of the new higher threshold.

(2)   An increase in child benefit to £10.35 per week. This would also need an increase in the basic rate of income tax to 46 per cent. Higher rates of benefit would involve abolishing the married man's allowance. Married couples without children do very badly under this option.

(3)   A graduated income tax (from 15 to 60 per cent) with increased child benefits, allowing £20 per week to be paid to those on the lowest incomes and £8 to those in the top bracket, but that top bracket would have to start much lower, at one and a quarter times average earnings. Seven per cent of families would lose more than £14 per week compared with the present state of affairs.

There seems to be no way of changing things so that no one is worse off.

A. B. Atkinson
'Shall we ever reform taxes and benefits?'
New Society, 16 June 1983

Other 'insulating' devices have been looked at, mostly to do with increasing child benefit allowance, which is the one insulated benefit which we have today. The point of all these devices is twofold: (a) to guarantee the means of a decent livelihood to everyone; (b) to encourage, not penalize, the first fruits of any initiatives to improve one's earnings.

The problem about implementing any change in the present system is that, given a fixed economic cake, any help for one group costs another group something. There is no device that is cost-free, as box 5.8 makes clear. Change in a democracy is never easy.

Insulation of benefits might, however, cost very little if it did indeed encourage more people progressively to work themselves out of the benefit system. All it would require is a change of attitude, a change from thinking of 'maintenance' to thinking of 'investment', from 'society's charity' to 'the individual's right'. We detect that attitude in child benefits; we see a lot of it in health care, to the degree that it is deemed by many to be antisocial to be independent; interestingly, we see it in education, where the benefit system is regarded as an investment by society – insulation is applied to the fees and to basic grants, and the Manpower Services Commission has astutely extended the principle to cover education in work on the Youth Training Scheme. Insulated benefits as temporary investments are, it seems, acceptable in certain areas (for children, the sick and the learners). Why could not the principle be extended to those who want to work? Why not regard all the unemployed as self-employed, deserving a salary until they are self-supporting?

More and more taxes

Try as we may to make more people self-funding and self-sufficient in the black, mauve and white economies, there is one unavoidable equation: there will be proportionately more people to be supported, whether they be young and in education of all sorts, old and in retirement or of all ages and unemployed. The calls on the state's funds seem bound to increase, whatever the avowed intentions of any Government. There will have in the end to be more tax collected, however hard a Government may try to find economies elsewhere, to sell its assets or to borrow.

The important question is not whether or even how much but how? The temptation for any Government must be to rely on income tax. It is easy to collect, progressive, easily altered and predictable. But in a time of increasing taxation and decreasing employment it is a slippery slope on which to step forward. There is, for one thing, the danger of creating a divided society. Higher income taxes will require higher incomes, leading to a mutually resentful society. Those in work will resent their high taxes, needed to support those not in work. Those not in jobs will resent the high incomes paid to those in jobs, paying little heed to the taxes the latter have to pay. Each side will resent the good fortune (as it sees it) of the other. Secondly, an income tax is a tax on labour, putting up the cost of production and therefore of the final goods. Higher taxes therefore mean lower competitiveness and lower production unless the number of jobs can be reduced still further. Either way higher taxes are in the end counter-productive – they lead to fewer income tax payers and less revenue.

The alternative to income tax is indirect taxation – VAT, excise duties and licence fees. But large increases in these are inflationary, because of their direct and immediate effect on the cost-of-living index, and regressive, in that they hit the poor proportionately harder than the rich. On the other hand, in a world where only half the adults are in jobs it could be argued that indirect or expenditure taxes are at least one way to ensure that eveybody contributes something, including the legally and illegally self-employed, the pensioners and those on state salary or benefits.

The choice is a tough cross for any Government. The chances are that we shall see more of each type of taxation – direct and indirect, income and expenditure. What would be misleading would be to pretend that any future scenario would involve less taxation, that faster growth, leaner administration or more oilfield bonanzas could lessen the burden. Even under a Government committed to reducing taxation and with booming oil revenues, Britain saw its percentage of taxation rise in the early 1980s. Governments may come to realize that inflation, if it is predictable, is not after all the greatest of all evils, that expenditure taxation does permit people some freedom of choice (they do not have to buy expensive luxury goods, or even go to the pub or the cinema), provided there is compensating provision in basic levels of (insulated) benefit. To comfort Governments in this difficult choice there is some evidence that Britons would be prepared to pay higher taxes to see a fairer society.

A probable first step would be to modify income tax and National Insurance taxes by turning them into an expenditure tax. This is not a tax on goods and services like VAT but a different way of measuring personal wealth. It taxes an individual's total *expenditure* rather than his or her total *income* and exempts all that goes into savings. Its main virtue is its simplicity – it would lump together cash received from capital sales and from wages, for example, but it would not necessarily raise more money or change the dependence on some form of income tax which, combined with National Insurance, accounted for 48 per cent of tax revenue in 1982.[5]

BOX 5.9   WOULD WE PAY MORE FOR OTHERS?

Many people in Britain in 1983 would have been prepared to pay more of their own income to see that everyone had such staples of life as a fridge, toys for children and a week's holiday away from home, as well as heating, proper clothing and adequate food. The existence of a widespread 'welfare consensus' was established by a survey done for London Weekend Television in 1983.

58 per cent would like to see higher pensions, while only 1 per cent thought pensions were too high.

59 per cent would increase supplementary benefit and only 3 per cent reduce it.

40 per cent would put up unemployment benefit and 9 per cent reduce it.

74 per cent would be willing to pay an extra 1p in income tax to provide the necessities for all, and

35 per cent would be prepared to pay 5p, or £5,000 million p.a. in total.

66 per cent would like to see the Government introduce a national minimum wage, and 74 per cent believe that the gap between rich and poor is too big.

Stewart Lansley and Stuart Weir
New Society, 25 August 1983

There is, however, one way in which Governments could increase tax yields without major tax reforms, and that is to cut the relief given to particular forms of savings and wealth creation – houses, pensions and life assurance. These have been called 'civil servant assets' because they are hard to convert into cash and are best suited to those with long and secure careers. They are also particularly unproductive forms of wealth, channelling savings into static reservoirs. In 1979 house, pensions and life insurance policies accounted for 68 per cent of personal wealth, whereas stock market securities (investment in industry and commerce) made up only 13 per cent. Twenty years ago the position was reversed, stock market securities accounting for 45 per cent and pensions, etc., for 40 per cent.[6]

The lost tax from this static wealth was estimated by the Government to be £4.3 billion in 1982–3. The tax gains that would result from abolishing relief on these forms of savings would obviously be less, since people would no longer divert as much of their income to houses, pensions and life assurance schemes. There are many who believe that this would do no harm at all; our houses are already too expensive, our pensions too fat and our lives too well protected, all draining money away from consumption and from investment in productive assets and thus ultimately from jobs and wealth creation.

These forms of tax relief also reward those in secure jobs as opposed to those in fringe employment or no employment at all. At a time of full and guaranteed employment this bias in the tax system could be justified on the basis that it was on the way to becoming a universal benefit and the only way for wage earners to accumulate capital. Today (and tomorrow) it looks more and more like perks for the privileged and a dangerous diversion of the nation's cash flows into static assets.

To abolish or reduce these tax exemptions would, however, effectively raise the rate of income tax for many, which would not only make the measure unpopular but would also add to rather than reduce the proportion of direct taxation. We should not, therefore, expect any major moves in these directions too soon.

Another step towards educating the public in the use of taxes may by 'hypothecation' or the principle of tailored taxes. A 'medical tax' would, for instance, be a tax collected to pay only for the National Health Service. It would not reduce our tax burden, but it would make it more obvious what the money was needed for. There are arguments against hypothecation – for example, it reduces the flexibility of the Government, and it makes it easier to spend money on popular and visible areas than on the unseen and unglamorous – but it may be the only politically feasible way to reduce general taxation while at the same time raising more money.

It seemed a simpler world when it could be assumed that every household had at least one job holder who would earn a wage, pay his income tax through the firm and in due course retire, often on an occupational pension at two-thirds of his final salary. The truth was never quite as simple. Assuredly, in the new world of work it will be more complicated still. Society has to use taxation to find a new approach to paying ourselves if we are not to create the divided society that might in time lead to the sort of revolution that Marx predicted. That could happen if the interests of the citizens, disappointed in their share of the wealth, turned away from work to attempts to gain control of the sources of power in the distribution of wealth.

Pensions, national income schemes, benefits and taxation policies are, in the end, very technical matters. But they ultimately work or do not work because of the philosophies which underlie them. The

Protestant ethic joined income and labour in line with St Paul's view that he who did not work should starve. That philosphy, however much it lingers on in the hearts and minds of many, cannot work where jobs cannot be guaranteed. A new philosophy is needed, on the lines that society supports its own, that every citizen deserves a livelihood as a right. But a right implies a converse duty – you cannot have one without the other. A 'right to work' implies a 'duty to employ'. When that becomes impossible for all, the right to work becomes a right to livelihood. Perhaps the converse duty is the duty to contribute, in some way, to one's society. *How* we pay ourselves is, in the end, to do with *why* we pay ourselves. It is a question that needs, urgently, wider debate.

WHAT DO WE USE FOR WEALTH?

The questions of what we mean by wealth, how much we need of it and how we create it are some of the more puzzling philosophical and technical problems of our time, all the more baffling because it looks as if their answers should be obvious and easy.

Thomas Mun, a rich English merchant in the early seventeenth century, saw wealth as gold and silver. To increase England's wealth he suggested that she must sell to other countries more than she bought from them; she must cultivate unused lands, be frugal in the use of natural resources, develop industries at home to supply necessities, avoid frivolous changes in fashion, reduce the consumption of foreign goods, carry English goods in English ships, cut prices where competition was strong, raise them where England had a monopoly, make England a centre for exhange, place no restriction on the export of money for overseas investment. As Tom Stonier comments when reporting this, it was 'not bad for early seventeenth-century thinking'.[7]

Many people would instinctively agree with Mun today. But his recommendations would restrict wealth to the balance-of-payments accounts – Britain's national housekeeping accounts. They would take no account of all the work that goes on in our hospitals, local councils, schools, police forces, etc., the cost of which gets added into the Gross National Product as part of the nation's wealth. Under this definition of wealth the more we do, the wealthier we are, even if the 'doing' means building more hospitals to put together the bodies mutilated on the motorways we build. This definition of wealth springs from the 'labour theory of value', first developed by Smith and later used by Marx, which holds that it is human labour which gives anything its value.

Once you go down that route, however, there is no reason to stop at the people who are actually paid for their labour in the formal economy. There are also all those in the informal economy, equal, as we have seen, to half the labour costs of the formal economy but not counted as part of the wealth of the nation. Or why stop at labour? What about talent? Is there not a sense in which Vienna in 1900 was rich in its artists, musicians, doctors, philosophers, beyond any normal concept of labour value? Or what about beauty, peace of mind, harmony and health? Britain may perform poorly on many indices of industrial output or economy efficiency but shows up well on rates of suicide, road accidents, longevity, murder and violence, crime and drug addiction. Which is more important? A 1977 Harris survey in the USA found that more people thought that 'finding more inner and personal rewards from the work people do' was important (64 per cent) than those who wanted to 'increase the productivity of our workforce' (26 per cent) or 'satisfying our needs for goods and services' (17 per cent). Surveys in New Zealand, Norway and France have come out with the same sorts of findings – non-material wealth matters more than material wealth to many people in affluent countries.

Where does all this get us? It explains, to start with, why any debate about wealth creation is confusing: people are using different definitions. To be explicit one needs to distinguish between the different kinds of wealth that are to do with *trading* (our ability to buy and sell in the international market place), *activity* (paid work, which is measurable and potentially taxable) and *well-being* (the less measurable elements of our quality of life). Any successful and civilized country needs all three varieties of wealth. Too little trading wealth means restrictions on what one can import, both food and raw materials (literally the 'stuff of life') and the consumer products which enrich our lives by providing more variety. Too little activity wealth means fewer jobs but also fewer incomes to be spent and to be taxed – incomes which are the spring from which all the revenues of the state and its services ultimately flow. Finally, too little of the wealth of well-being would mean a society of workaholics, totally taken up with the business of survival and of achievement in the dogfights of human ambition. If the future world of work is going to be worth living in, we need as much as we can get of all three kinds of wealth.

Trading wealth

Trading wealth depends on (a) our ability to make or do things better or cheaper than others so that others want to buy them;

(b) our ability to make things for ourselves at least as well and as cheaply as others can, so that we do not need to buy from others. Both (a) and (b) are important to our trading wealth. If we ourselves made more video machines which were at least as good and as cheap as the imported models, we would save money on imports and increase our trading wealth, even if we did not sell one of them abroad. Import substitution is just as important as exporting. North Sea Oil is both at the same time, which is a fortunate convenience but a relatively temporary one. Oil, like time, runs out. Fortunately, there is still enough of that time left to take specific measures to increase the different types of wealth.

We have to grow new businesses which do things that others cannot do as well. For Britain this has to mean the 'know-how' services and goods; knowledge and applied technology could provide a competitive edge, even if they are dearly priced. Britain has long been renowned for its know-how services, which include not only finance, insurance and banking but also television, women's fashion, the arts, consultancy. There is also a long tradition of superior products which depend on original know-how for their commercial success rather than the cheapness of their production. Britain, no matter how hard she tries to improve her productivity, will never be a cheap producer by international standards; her interests, therefore, will be best served by consistently finding and making the new, letting others do the making once they are old and established. Britain's trading wealth depends on her being the laboratory and the test bed of the world, not its assembly shed. This is already happening to some extent (see box 5.10) but not enough for anyone's comfort.

No one can legislate new know-how businesses into existence, but more could be done to create the right conditions so that what seeds there are will germinate. More could be done too to produce more seeds. Higher education ought, for instance, to be a national priority area, particularly in the sciences. In 1981 9.3 per cent of the boys leaving school went on to a degree course of any sort and 6.8 per cent of the girls. Only 15 per cent of those would be studying science and 18 per cent engineering and technology. Add those together and you realize that less than 3 per cent of our young people were studying for a degree in science or engineering in 1981.[8] By international standards that figure is puny. Higher education in Britain is very selective, turning away many potentially good candidates. It aims to educate the cream. But, unfortunately, a lot of cream requires a lot of milk to start with. Quantity in education is a prerequisite of quality because no one can predict with certainty who will excel at each stage and who will not. The strategy for higher education in

BOX 5.10   THE SEED BEDS OF THE FUTURE?

Cambridge now has more than 400 electronics companies, at least ten companies making micro-computers and the largest of the 'science parks' in Britain – a nursery for new businesses closely linked to, and partly funded by, the university. Some of the new impetus comes from young Cambridge companies which have encouraged their employees to found their own businesses, while bank managers in Cambridge have competed to lend to the new firms in a most un-British banking way.

Aston and Heriot-Watt are copying the 'science park' idea, but slowly. Aston's first tenant moved in only in 1983. Heriot-Watt has five. Other universities, including Salford, do contract work for businesses, and Salford staff and students have actually set up a new business of their own to make a new kind of alarm.

The Scottish Development Agency has encouraged multinational enterprises to spawn new businesses. They can count sixteen new electronics firms in the past four years, founded by people from the multinationals. Rodine, making disc drives for computers, was formed by people from Burroughs, Fortronic by people from Hewlett-Packard, and they in turn provided the founders of Subscan.

In Berkshire, the oldest electronic centre in Britain, hundreds of new companies in the electronics and information industries have sprouted up in the last five years.

. . . and the propagators?
A wide variety of institutions are now interested in providing money for new ventures. This venture capital industry includes the following.

Venture capital funds – which raise money from the institutions and sift out new proposals. These funds include old-established ones like Newmarket and Abingworth as well as newcomers such as Advent and Alta-Berkeley.

Merchant banks – which have traditionally been a source of advice and risk capital but of which some are now taking a special interest in new technology businesses. Rothschild, for instance, has a biotechnology investment fund.

Insurance companies – the Prudential, Britain's largest insurance company has set up Prutec to handle its venture capital interests, while the Commercial Union and Legal and General have started Logent.

Management consultants – PA raised £6 million for a fund called Managed Technology Investors.

BOX 5.10 continued

The Government — the Government gives an 80 per cent guarantee on loans to small firms and passed, in 1979–83, more than seventy measures to support small businesses.

In the period 1980–3 the Government guaranteed loans worth more than £300 million, and over £200 million of venture capital was raised.

Economist, 23 July 1983

Britain is a statistical gamble, relying on unusually good results from unusually small numbers. It is also unfair to those who have not learned enough, or been taught well enough, by the age of 18 to jump examination hurdles better than their contemporaries. To ration higher education as severely as Britain does today is to run a lottery on our future.

New businesses need not only talent but also time and room to develop, and they need co-operative customers in the early stages. They need patrons, in other words, as talent always has throughout the ages. Some of those patrons will be the banks, venture capital funds and big businesses of the type cited in box 5.10. But the principal patron in the high-technology stakes has to be government; government as sponsor of research, government as commissioning customer, government as co-experimenter. Wars have always been great generators of new products, mainly because in wartime governments inject a new urgency and emphasis into their activities as patrons. It would be dangerously naive to think that private business is sufficiently rich and farsighted to assume all the patronage on its own. More and more governments have realized this, as box 5.11 demonstrates. In Britain government still looks to the universities to be its research laboratories, backed by the British Technology Group, charged with the development of their results, but the Ministry of Defence has realized that its own establishments may be sitting on marketable inventions and has licensed groups of entrepreneurs with access to venture capital to go hunting within R. and D. establishments. Britain is not large enough to indulge in the shot-gun approach adopted in the USA, which involves a profligate use of resources, nor is she organized and disciplined enough as a society to emulate the laser-like precision of Japanese industrial planning. Like all patrons, government has to combine encouragement for all and the selection of a few.

Governments, however, should not have a monopoly on patronage. Big business, which itself sprang from small beginnings, could also think of appropriate forms of patronage. Many of the mushrooming companies of America's Silicon Valley and Route 128

BOX 5.11   GOVERNMENTS AS RESEARCHERS

European governments have increasingly recognized that in some sectors of industry only government can afford to pay for the research required. Examples are:

France A £3 billion per year programme in telematiques, including the provision of free viewdata terminals to all telephone subscribers.

Italy The Instituto Mobiliere Italiano (IMI) has been set up to encourage and sponsor R. and D. projects in Italian industry. By 1977 £1.5 billion had been committed.

Holland In the late 1970 Holland began a programme to spend £90 million a year to encourage R. and D.

Switzerland The Swiss have set up a programme to encourage the rapid development of electronics technology and to sponsor applied research and development.

West Germany The West Germans have a variety of support measures to aid R. and D., including subsidies, tax relief, venture capital, low-interest loans, all supervised by the Ministry of Technology in close liaison with the Ministries of Economics and Education.

Tom Stonier
The Wealth of Information: A Profile of the Post Industrial Economy (Methuen, London, 1983)

were started by individuals, leaving the big corporations to pilot their own ideas. Rather than resenting this abandonment, the big corporations should perhaps encourage these new seeds of enterprise, seeing themselves as the breeding grounds of inventors and entrepreneurs, even if these have to leave the corporate grounds in order to experiment. It is not ideas or talent that these aspiring individuals lack but finance and access to markets. Here the corporation can help, perhaps with advantage to itself but certainly with advantage to society. The merchants and the kings who patronized the creators and the artists in earlier ages did it in order to promote good public relations and out of a sense of social responsibility, not personal gain. It would be appropriate if the corporate princes of the last industrial age were to be the patrons of the new one through their sponsorship of talent.

Growing more know-how businesses will improve our trading wealth by improving our capacity to export. This form of wealth comes, as Tom Stonier points out in his book *The Wealth of*

Information, by turning a non-resource into a resource by intelligence and diligence. Silicon was valueless until it became the base material for chips. The floor of the North Sea was a non-resource until applied science turned its fossils into the wealth of oil. Stonier lists other possible areas in which intelligence and diligence could create wealth where before there was nothing. His list includes: waves as a source of energy, coastal fish farms, deep-sea farming and mining, photovoltaic cells to create electricity, forest farms of food-producing trees. Some of these notions may be fantasies, but there will be other lists as long as clever people try hard to devise them. As Arthur Scargill once said of Britain, how can a land that is built on coal, surrounded by fish and with more waves per head of the population than any other country ever be poor? Only if we do not provide enough clever people with enough time and enough encouragement.

Activity wealth

More people must be encouraged to make and do things for the local and national markets. The internal economy must be stimulated, whether or not more gets exported. More small businesses may not solve unemployment, but they will mean more work for some, more taxable flows of money, more alternatives to imported goods.

How is this to happen? The lure of the market place does not seem to be sufficient in itself to call out the entrepreneurial energies of the British, even after taxes have been shaved at the top and a whole medicine chest of drugs and stimulants put on offer for small businesses.

Once again, patrons are needed to get things started, to prime the pump and to take some of the risk out of the decision to start or expand a business. Preferably they should be local patrons because small firms and new firms do not stand tall enough to see the national scene.

What form might this patronage take? There are several possibilities.

First, local enterprise trusts. These are consortia of local businesses which offer advice and facilities to new businesses in their area. It is a way in which larger employers can put some of their experience and contacts back into the community. There are now over 100 of these consortia, some more practical than others, for it is temptingly easy to join one of these groups, less easy to give the kind of practical help that a new enterprise needs, which tends to be cash, materials, premises or customers rather than advice.

Second, work centres. The one thing that every new business

needs is space, be it a drawing board for a designer, a desk and computer for an accountant or a small shed for a manufacturer. The space is often there, lying derelict in our inner cities or locked up behind the gates of closed-down factories. Increasingly these spaces are being rehabilitated to be work communities. Covent Garden in London is the most obvious example of an urban space turned over to small enterprise in an imaginative way, with a noticeable increase in activity wealth and, probably, in the wealth of well-being as well. Job Creation Limited is another fruitful offspring of unemployment, which started by trying to create small businesses but now concentrates on providing the space in which they can come to birth – a new-style patron of enterprise.

A third, and more dramatic, act of patronage would be a major revival of the housing programme, pump-primed by the Government but funnelled through local authorities into the local business community. Giles Merritt, in his book *World Out of Work*,[9] recommends this as a major strategy for reducing unemployment, pointing out that if Britain were to build 500,000 homes a year (which are needed to update our housing stock and to provide an adequate 'float' of houses to encourage mobility and a freer market in property), then 900,000 people might be pulled into employment. It would cost £9 billion perhaps, partially offset by £5 billion saved in unemployment costs, but it would be an enormous addition to activity wealth, one which, moreover, would be largely non-inflationary and would not leak into imports because it would mostly go to the smaller, more local builders and to UK suppliers (see box 5.12).

The scale of Merritt's plan may be over-ambitious, even if the direction is right. Of all the possible infrastructure investments pressed on the Government by Keynesian expansionists, this is the one which perhaps could do most to meet a social need while at the same time stimulating new business activity, activity which in turn will generate more activity as those back in work spend their money – true activity wealth. Unfortunately, a speculative housing programme is unlikely to be given high priority by any Government committed to reducing public expenditure.

Fourth, more patrons are needed. Building societies are barred by their structure from lending to local businessmen whom they know. Banks are still more interested in covering their risks when lending money than in stimulating long-term growth. Business executives, who should be the new entrepreneurs, find it more profitable to invest in houses or pensions at tax-free rates, rather than to build new businesses. Solicitors, who used to finance small business in Victorian times, are hedged about by restrictive practices. Big

BOX 5.12   HOUSING – AN OPPORTUNITY FOR PUMP-PRIMING?

Every year in Britain 180,000 new households are formed, but only 150,000 new houses are started (97,000 privately).

Every year at least 1 per cent of Britain's 20 million dwellings need replacing – another 200,000 houses.

If mobility of labour is to be a real possibility, there probably needs to be a 10 per cent 'vacancy reserve' of available homes instead of the current 5 per cent – another 2 million homes, another 200,000 over ten years.

Seventy per cent of Britain's occupiers would like to be owner-occupiers. Only 50 per cent are.

In 1971 50,000 *more* houses could have been started if local authorities had been quicker to release land.

Every new house will provide two and a half jobs a year if those directly and indirectly involved are included (at no cost to government as far as private housing is concerned).

The Dutch are planning to put up 1 million net new homes, by 1990, proportionately five times the rate planned for Britain.

Adapted from the argument present by Giles Merritt in *World Out of Work* (Collins, London, 1983)

business, which could act as banker to fledgling enterprises, is too busy with its own concerns and finds it more convenient to delegate this reponsibility to a local enterprise trust. The Civil Service, which could use its patronage to encourage new enterprises, finds it safer to rely on tried and trusted performers. Aunt Agatha, the legendary provider of the first £1,000 of equity, no longer has £1,000. The land distribution programme, which gave so many Japanese entrepreneurs the assets against which to borrow, has gone the other way in Britain, into the hands of large farmers or commercial conglomerates. Pension funds, which one might think ought to be investing in Britain's future, are precluded by statute from risky investments, although a few, like the Prudential, have found ways around that embargo. More would-be patrons need to follow its example, for activity wealth is not self-generating.

The wealth of well-being

We all need money, but money is not the whole of wealth. What, then, is the rest? James Robertson, in his visionary book *The Sane*

Alternative, quotes with approval the words of Tom Burke, formerly Director of Friends of the Earth:

> The new wealth might count as affluent the person who possessed the necessary equipment to make the best use of natural energy flows to heat a home or warm water. The symbols of this kind of wealth would not be new cars, TVs or whatever, although they would be just as tangible and just as visible. They would be insulated walls, solar panels or a heat pump.
>
> Wealth might take the form of . . . access to enough land to grow a proportion of one's food. This would reduce the need to earn an even larger income in order to pay for increasingly expensive food. Wealth would consist of having access to most goods and services within easy walking or cycling distance of home, thus reducing the need to spend more time earning more money to pay for more expensive transport. A high income would be less a sign of wealth than of poverty, since it would indicate dependence on a job or a workplace in order to earn the income to rent services. Wealth would consist in having more control over the decisions that affected well-being and in having time to exercise that wealth.[10]

A sense of independence is one aspect of well-being. So is the opportunity to be more than someone who sleeps to eat to work to live. There is the well-being that comes from a rounded life, with access to sport, recreation, friends and community, the opportunity to share in activities, to contribute as well as to receive, to give and to get without the complication of money. *Inter*dependence, in other words, is as important to most people as *in*dependence.

The steps towards a greater wealth of well-being are difficult and more intangible than in the case of the other forms of wealth. They are three.

First, an attitude of mind which consciously values the lack of dependence and puts freedom above income in certain cases. This involves a cultural revolution in the definition of wealth, but there are some signs that we may be entering a period of change in cultural stereotypes. Independence and community are both in vogue, particularly among the young, with money taken for granted rather than esteemed. It is a value system which fits well with new forms of organization and, indeed, with new patterns of working life, but it fits badly with economic theories built on the assumption that mankind is forever greedy.

Second, conscious enabling. Ivan Illich has pointed out that many of the occupations in our societies have a vested interest in making people dependent on them, on disabling their clients.[11] Doctors need sick people; oil companies need oil-consuming customers. The sicker and the more oil-consuming, the better for their providers, financially at least. And so it goes on. Plumbers need people who cannot repair their own pipes; professors need students who need degrees. There is a curious unspoken conspiracy in much of society: half of us are glad to be incompetent and on the receiving end, and half are happy to assure the rest that we are indeed incompetent and need their services or their products.

The wealth of well-being requires this process to be reversed. Doctors should be interested in healthy people not sick people. Oil companies should help us to be less dependent on oil. Plumbers should help us to learn home maintenance and professors should instruct us to learn at home, without degrees. It is hard, at first, to see what is in it for the doctor or the oil companies who have thrived on disabled customers. But James Robertson points out[12] that there is a precedent in the dissolution of the old British Empire, when the old colonialists took a positive pride in developing the capabilities of those very populations they had for so long believed, and kept, incapable. The cynic, however, might say that the hand of the British was forced by circumstances, by the clamour of the peoples for independence. So it may be with us if we clamour hard enough for help to stay healthy rather than to be healed, to be self-sufficient rather than dependent and start to support those new professions and businesses which dedicate themselves to equipping us to be independent of them.

Third, community development. Independence is only part of the story. Interdependence with some sort of community is another. We need consciously to foster local communities. Community arts and recreations are one kind of focus. They need their patrons, who cannot always be central bodies like the Arts Council, whose funds can only help a finite number of communities. Business and local authorities, both hard-pressed always, must expect to find yet another call on their patronage – which does not always have to take the form of cash. People, premises and products can be equally useful at times. Indeed, it is people who provide the core of the networks which are most needed in most communities. Some of these people could be loaned by local employers on secondment or part-time release or as part of a corporate team project. This kind of community involvement is a growing phenomenon among American businesses. Over 500 corporations now have some kind of formal release programmes to allow their employees to become involved in the community.[13] An initiative by the Volunteer Centre in England is trying to encourage similar initiatives this side of the Atlantic.

There is evidence that many people, particularly those who are unemployed, would like to offer their time and talents to help the community.[14] What is needed is more structures for involvement. It is a small investment with a potentially big pay-off in this less tangible but still very real form of wealth, the new wealth of a new society. They could be paid, these volunteers, not in cash but in professional and personal development. If individuals saw the voluntary sector as a way not only of contributing to society but also of acquiring for themselves some of the experience and credentials of a professional in areas such as counselling, welfare work, housing or child-care, volunteering would grow even closer to 'real work'.

WHO PROTECTS US?

More small businesses, more self-employed people, a growing black economy, more do-it-yourself – it can sound like more freedom but it also sounds like anarchy or the absence of rule. Who, for instance, is going to make sure that the products of the small new businesses are as safe as those of the big corporations or do what their makers say they will? Who, in short, is going to protect the consumer in this free-form world?

Who, on the other side, is going to protect the individual at work, see that he or she is properly paid, properly treated and properly protected by law, insurance and pensions? Is each one of us to be his or her own protector? Are we to negotiate our own fees and contracts? Are we each to bear the consequences of any imprudent purchase? How are we to know that the plumber who calls at the door is any good as a plumber or that his fee is appropriate? How is he to convince us?

The large employment organization has protected us from many of these dilemmas for so long that we have tended to forget that they exist. It is neither easy nor circumspect for a large corporation to cheat on the weight, the quality or the size of its goods. The scale of its operations makes dishonest practices difficult to sustain and very liable to be discovered by someone somewhere. The development of collective bargaining by agreement between employers and unions has protected most of us from having to do any more than complain about our pay. What shall we do without the protection of those large institutions?

The question is easier to ask than to answer, but there are some pointers from the present and the past.

Agents

Self-employed people in the arts, publishing or the media have traditionally used agents to find them business and to negotiate terms. Why should not more of the self-employed have their agents? (see box 5.13.)

BOX 5.1   DO HOMEWORKERS NEED AGENTS?

There are estimated to be between 100,000 and 400,000 home-workers in Britain today. Councils are supposed to keep registers, but they know only if employers tell them. Not all do.

In 1981 Andrea Waind compiled a list of horror stories (*New Society*, 1 March 1982). Jan Smith was paid 24p *per day* for making up handbags. Post-machining, stitching linings to shoe uppers, pays better – £50–£60 per week – but no holidays and no sick pay.

In Leicester, the centre of hosiery outworking, the average pay in 1983 was between 50p and 90p an hour (the Wages Council minimum being £1.50). Some 10 per cent were earning 15p an hour.

The 'Fagin Spot' in the magazine *Outworkers' Own* featured a firm in London paying Leicester outworkers £5 for the seventy-two hours it takes to knit an Arran jumper.

Co-operatives

Producer co-operatives have had a chequered but usually sad history in Britain, but marketing co-operatives have a more successful record. The problem for many a self-employed person is not how to make the goods or deliver the services but how to find customers for them. Just as craftsmen in days gone by would group themselves together in streets or areas so that would-be customers could find them easily, and just as the new craftsmen congregate in refurbished warehouses, so more and more self-employed people will find co-operative structures to market their wares. Nearly 500 small co-operatives were set up in Britain in 1982, involving over 5,000 people.[15]

Guilds and associations

The equivalent of the unions for the self-employed or the professional was a guild or an association. These provided, and still provide, certification and regulation of fees. The principle needs to be rediscovered and extended so that the gardening expert, the computer programmer, the plumber and the carpenter can all be seen to be licensed to practise, with agreed scales of fees. An association by itself is not enough. Certification, self-regulation and public accountability are essential. Without these an association is only a club for the mutual comfort and protection of its members – a monopoly without accountability. As Lloyd's of London discovered in the insurance world, a self-protecting club is not judged to be in the public interest.

Unions

Unions will remain both the agents of collective bargaining in large corporations and advisers and advocates for individuals. We may expect to see their advertising and advocacy roles become more important as collective bargaining declines in line with the decline in large plants and employment generally. Unions, in other words, will begin to move back towards the guilds but will retain an agency role unless they add training and certification of individuals to their representative function. There are no signs that they accept this kind of future as yet. In common with many organizations facing decline, they will tend to work harder at what they have always done rather than face up to the possibilities of a new future.

Consumer education

A freer and more anarchic market will put more pressure on the individual consumer to decide what he or she wants and to check whether the product lives up to its specification. This is an invitation for more literature of the *Which?* variety, more standard-setting institutions on the model of the British Standards, more investigative reporting like that of the BBC's *Checkpoint* and a growing Office of Fair Trading and other mechanisms for complaint and redress. A more open and more competitive market may in the end be best for the consumer by sorting out the poorer products, but the sorting out can be a lengthy and, for some, an expensive process.

Mutual help groups

A more dispersed and free-form society will offer more holes for people to fall through. The society that is to come will, one must hope, have its caring side, expressed in the social services of the state and in the local communities, but it will also be, for many, a lonely and competitive world in which those who do not get on get left out. Mutual-help groups are the co-operatives of the misfortunate. The poor, the lonely, the sick and the handicapped, the old and the abandoned may find it hard to make their way around a world which will look increasingly like a vast library of Yellow Page directories. They will need each other, and every encouragement should be given to them to form their own co-operatives. Such groups cost no money and require few premises but can offer very practical help and comfort. To help them on their way more information is needed about where to look for assistance. A computerized list of networks called up on the television set is one possibility but probably for some time beyond the reach of the groups for whom it would be most help, unless we follow the French and provide a computer with the telephone. This is intended as a way of making the population comfortable with the computer. It may also be the best way of linking people into networks.

BOX 5.14 A COMMUNITY WORK STRUCTURE: THE CUMBRIAN SKILLS-SWAP NETWORK

Walter Goult is a mechanical engineer who was made redundant and decided to use his free time for the benefit of the local community by organizing a swap shop for skills.

After 16 months the scheme had 370 members and 98 skills over a 30-mile stretch of Cumbrian valley. There are nine meeting centres where plumbing skills can be swapped for gardening, care and social links formed over tea or a drink, or outings and parties. No money changes hands, and skills are not evaluated or 'exchanged' but given for free, so no benefit is lost by those who are unemployed. The result is a net increase in living standards and well-being by all concerned. It is a work *and* a social network.

It is also a place of learning. Young people can ask to be taken on as assistants by a bricklayer, swapping their labour for a bit of learning by apprenticeship. There are plans to raise money for a furniture-repair workshop which will make more training possible – again for free.

With a little initiative a community has created a structure for itself which provides work, social contact and learning without costing anyone anything except their time.

Guy Dauncey
'Nice Work if You Can Get It', National Extension College, 1983 (Other self-help groups, particularly for the unemployed, are listed in Dauncey's work.)

It has already taken too many pages to etch out some of the possible answers to just four of the questions on the new agenda posed by the new future of work. The final answers need much more technical discussion, much more research and experimentation, than they have yet had, or can be given, in this book. That is, however, the best argument for giving them a wider airing in all quarters. In November 1983 the Joseph Rowntree Memorial Trust organized in Yorkshire a talkabout on the future of work, with the intention of increasing local awareness of the issues involved. A national talkabout on this agenda is long overdue.

It is understandable that politicians should shy away from it. This is one area in which no one can promise cake for free to all. None of the options sketched out in this chapter is free. Someone has to pay, and if you add them all together, someone will be paying quite a lot. More taxes seem inevitable. Longer lives and shorter jobs will tend to mean lower incomes. Some no doubt will get richer, but most will feel poorer, financially, at least for a while. The new world may have more wealth of well-being but, for most people, less conventional wealth unless we are very clever or very lucky. Sacrifice, therefore, is inevitable. In a just society that sacrifice will be shared. In a jungle society the sacrifice could be lumpy. If we don't want lumpy justice and lumpy sacrifice, we need to talk about the questions arising, and soon.

6

Educating for Tomorrow

I found myself recently explaining to a conference of head teachers that education will be the one guaranteed growth sector of our society in the next thirty years. This was an odd thing to say on that occasion because the conference was about the problems of contraction in the education service due to falling numbers of students.

It depends, of course, on what you mean by education. Education is being redefined very fast, and many more people and institutions are getting in on the act. The schools may be contracting, but the educational work of the Manpower Services Commission is exploding. The first year of the Youth Training Scheme, in 1983–4, aimed to take on enough young people to fill fifty medium-sized comprehensive schools. Then there is the education that is sold as a leisure interest – the books on travel, gardening, house maintenance, history and cookery that fill the high street stationers. Television sets and video recorders have put a self-administered classroom into many a house; learning dressed up as entertainment has become a popular good. The home computer starts by being a personal Space Invaders console but can become an exotic, and private, way to learn mathematics or typing. Fifty per cent of all adults would like to take a course in something, even if only half of them may actually get around to it. The universities and polytechnics are over-subscribed; more and more people want to stay on at school after 16 – not just to keep off the streets but to invest in their own development.

This is no accident, of course. It is all part of the changing scene. More people have more discretionary time. More people need to equip themselves with useful and saleable skills. Investing in yourself is making more and more sense in a world where you may have only yourself to depend on.

However, just as the new education is booming, the old education is contracting. Public spending cuts have meant that universities and polytechnics have been asked to cut back just when the full effects of the baby boom of the 1960s is arriving at their gates. Schools, hit by the dip in the birth rate which started five years later and by the cuts, are having to contract, close or merge. That is not the best atmosphere in which to experiment, to adapt to a new world of work and to a world with different values. Nor is it particularly helpful to be reminded that while you are struggling with fewer resources, the world around you is bursting with a demand for learning and a bevy of innovators. The temptation for a besieged monopoly is always to retreat into doing what it knows best how to do, to reinforce its central strong point and to shut its gates to the outside.

We are at a critical point in the history of education. On the one hand, it is clear to many people that we need more education, with greater variety, for more people at more stages of their lives than ever before if we are to make a success of the new world of work and leisure. On the other hand, we have a formal educational system that has traditionally been turned in on itself and, under the pressures of contraction, is likely to become more so, teaching people what it can teach them rather than what they need to know, preparing them at each stage for the next stage but not necessarily for life itself.

Can we, therefore, trust the formal system to deliver what we want? Can it change? Or should it be left to wither, concentrating only on what it can do, while other institutions grow up to meet other needs? (The British, I sometimes feel, do not change their institutions, nor do they kill them. They simply turn their backs on them and start something else.)

The question is crucial, and there is not much time. The individuals who will be in charge of society in 2020 are already in school. What kind of educational future will they face if we do not change things? Come to that, what educational future will their parents face, for in spite of the tradition that schooling finishes in one's teens, learning goes on through life? In this chapter we shall look at the problems of the present system, at the potential for changing it, at the need for more serious educational provision in life after school and at the kinds of actions which are needed to make this happen. But first we need to be clear about why education is going to be so important, even if it takes new forms. There are six reasons.

(1)   The creation of new trading wealth depends on the positioning of some people at the frontiers of science and the new technology. To have enough of this cream we need quite a lot of milk to start with, as the last chapter emphasized.

(2)   A changing technology will require skills and the ability to learn new things and new habits two or three times in a lifetime.

(3)   The growth of the services, of the professions and management

and of personal services will also mean the growth of what has been called the 'credential society', in which most people will need a certificate.[1] Education for credentials will be increasingly important.

(4) More of us will have more time to do things for ourselves in our free time. Education improves our capacity to find something useful and pleasurable to do. In the last analysis it is better to be educated and unemployed than uneducated and unemployed.

(5) Education is itself a great creator of employment. It is one of the most labour-intensive of the service industries and will probably continue to be so, in spite of new technological aids. The Open University, with its battery of distance learning tools, still needs its tutors.

(6) Education as such is a 'good thing' and the mark of a civilized society. The wealth of well-being requires that we recognize the work of knowledge, of expanded consciousness, capability and creativity, as well as of material consumption. *Active* pastimes and sports, doing not watching, seem to be closely related to educational level. Amateur dramatics, birdwatching, music making and hang gliding are not the pursuits of the so-called 'non-examinables' at school.

A DISABLING SYSTEM?

The British educational system today probably harms more people than it helps. That is not intentional. The teaching profession is, on the whole, both diligent and dedicated. It is the fault of the system, designed at other times for other purposes but now disabling rather than enabling to many who pass through it. In 1981, as proudly recorded in *Social Trends* 1983, 51 per cent of boys and 56 per cent of girls left school with at least one 'O' level or CSE Grade 1. The other side of that coin is the shameful truth – 49 per cent of all boys and 44 per cent of all girls left school as failures, without a single worthwhile certificate to their names. The great bulk of them become the 40 per cent of our youth who recieve *no further training at all* in life after the age of 16 – 40 per cent of the nation written off educationally.

How has it turned out that way? Through a combination of circumstances, each of them well-intentioned but with unintended consequences. Consider the following points.

(1) Compulsory schooling was originally introduced to keep young children out of the mines and factories.

Unintended consequence: education came to be seen as an alternative to work rather than a preparation for it, an education for civilization rather than for industry.

(2) In 1917 a Government harassed in time of war, by administrative Act, made the universities responsible for administering school-leaving examinations.
Unintended consequence: universities applied the only criteria they knew, those which they used for their own entrance examinations, knowledge and analysis, to *all* pupils, irrespective of whether or not they wanted to enter university.

(3) In the 1960s and 1970s a revulsion against selection at 11+ led to the establishment of comprehensive schools in which all children had access to a wide range of subjects.
Unintended consequence: all students were now directly exposed to the common criterion of knowledge-based examinations. 'Mock' exams at 15+ took the place of 11+ tests, effectively dividing pupils into 'O' level material or not.

(4) In a desire to make examinations fair, the marks were adjusted around the average for each year (norm-referenced) with a certain percentage getting As, another percentage Bs and so on.
Unintended consequence: no matter how well people do, a certain percentage is bound to fail.*

(5) In a desire to have one uniform graded system of leaving examination, the 16+ GCE scheme was designed, to suit all pupils at all levels of ability, giving them an 'O' level or a CSE.
Unintended consequence: employers have only one criterion to use in judging applications so go for the top of the system, thus reinforcing the primacy of knowledge and analysis in the assessment of people's potential.

(6) In order to improve the standards of teaching, the teaching profession was increasingly professionalized, requiring eventually the equivalent of a degree followed by a year of educational training.
Unintended consequence: new entrants joined the profession straight from school or college with no experience of the world outside education, thus reinforcing the view of education as something apart from work.

---

*In January 1984 Sir Keith Joseph, the Secretary of State for Education, in a speech in Sheffield, indicated his wish to see some more objective tests replace norm-referenced examinations.

(7)   In order to save public money, and improve standards, the grants to Universities were cut in 1980–3.

Unintended consequence: since this coincided with the baby boom leaving school, there was a big rise in the standards required for entry, which put even more emphasis on the examination system as the ultimate record of achievement.

The consequences of these well intentioned reforms were foreseen by some. In 1959 the Crowther Report said:

External examinations not only tend to direct attention, and attach value, to the subjects which are examined at the expense of those which are not . . . they also focus attention on pupils who are examined at the expense of those who are not. Many, probably more than half, of the pupils of the modern schools would have their education deflected from its proper lines by being prepared for an examination.

Twenty years later Peter Wilby, the education correspondent of the *Sunday Times*, commented:

The trouble with our comprehensives is not that their academic standards are too low but that they are too high . . . Our secondary education is organized to select those few who will go to university and, ultimately, the even tinier minority who will approach the frontiers of theoretical knowledge. For their sake, all our children are being put through an over-blown, over-academic syllabus in which the dominant experience, for the majority, is one of failure, not of achievement.[2]

It is not a new phenomenom either – see box 6.1.

But it is not just the examinations. To be a pupil in a large school is a strange experience. How many of us, if asked to organize an office, would so arrange things that people worked for eight or nine bosses in a week, in perhaps five different work groups, in seven different rooms, without any desk or chair to call their own or put their belongings and discouraged, if not prohibited, from talking to anyone while working? Furthermore, which of us would then interupt them thirty minutes into each task and move them on to the next? Only slightly caricatured, that is the experience of a pupil in a large secondary school.

The truth is that, organizationally, the secondary school is not organized around the pupil as *worker* but around the pupil as *product*.

BOX 6.1   SCHOOL EIGHTY YEARS AGO

The week's curriculum, I fancy, must have been a bit of a conundrum to all concerned. Why were we learning or not learning these things? Of course, reading, writing, arithmetic have their own sense to them ... But drawing, but history, but grammar, botany, singing, geography, geometry, recitation – what mortal use could they be to the likes of us? Our parents said, to help us get better jobs. Our teacher said it was a fine thing to be educated ... The fact is, not they nor anybody could say plainly what we were being educated for.

In a few schools up and down the country, teaching is a simple matter because the pupils have a reasonably foreseeable future which can be contemplated cheerfully. They have waiting for them the same assured position in adult society that their parents had. The teacher must prepare them for that position by the appropriate character conditioning, initiation into the peculiar code of behaviour which is the mark of their kind, and a laying on of the gold leaf of culture to make them look worthy of the job already picked for them.

But our lot of kids were just going to be ordinary workers ... Now what character, what code of behaviour, what culture is appropriate to the worker?

They hold revolutions about that in some places. You couldn't expect our teachers to have the answer ... Our school was doing well it considered, by the only practical test that existed, if it managed to raise the proportion of pupils capable of winning scholarships and getting there by possible passports to Better Things ... Always the pride that prevailed in this working-class school was that it succeeded in turning out less recruits for the working class than any other of its kind in the district. That less was still the majority, mind you...

But the school's official boast was not of them.

Jack Common
Kiddar's Luck (quoted by David Hargreaves in *The Challenge for the Comprehensive School*, Routledge & Kegan Paul, London, 1982)

Raw material is passed from work station to work station, there to be stamped or worked on by a different specialist, graded at the end and sorted into appropriate categories for distribution. They do things differently in primary schools and sixth-form colleges, but the secondary school is the definitive sorting mechanism, and it leaves an indelible impression. That so many come through it, smiling, grateful and grown up, is a tribute to the dedication of many teachers who impose their humanity and personality on those huge processing-plants. But many do not come through as well. They leave alienated by an institution that seems to them oppressive, irrelevant and dismissive of their possible contribution to the world. Truly, for them, it is a disabling system.

Why does it have to be so?

Why does there have to be one national leaving examination at 16+? No other country (apart from the USSR) has one at that age. Why could not each school graduate its own students (as in the USA)? Or why could they not take a proficiency test in each subject when they were ready for it, at the appropriate grade, as in music, so that you would leave school with Grade 6 in this subject, Grade 4 in that, Grade 1 in the other.

Why, in fact, do we have an assessment system so biased towards one dimension, as if we measured every flower by the size of its petal or every book by its length? Why could there not be more multi-dimensional assessment, more tests in which 'clever' did not always mean 'best'?

Why too is age so important a factor? Why do we do everything in schools in age bands? Why do all 14-year-olds have to do everything together? It is not a condition that we impose on adults, even on people wishing to marry each other. We advocate mixed ability and a common age: why not common ability and mixed ages?

Why is co-operation (even called 'cheating') so frowned upon in education when *work* is always organized around co-operating, sharing groups? Do schools falsely encourage an individualist approach to work because that suits their examination system?

Why do teachers have to be so specialized that they can only teach one subject, no matter how advanced or how elementary their pupils? Is not the *process* of teaching at least as important as the *content*? Could not most degree-bearing teachers teach a range of subjects to children under 16?

Why is it so important to have schools covering the age range 12–18 and not 12–16 or even 12–20? The insistence on a sixth form with at least ten subjects requires a huge entry at the bottom of the school to generate large enough classes in enough subjects later and encourages teachers to specialize. Size and specification together result in the process-plant organization so imimical to learning. No one in manufacturing today would think of building a plant to house as many people as the average comprehensive school.

Why does all learning have to take place in schools, which are best at teaching those things which can be taught to large groups in small rooms – like knowledge – but cannot simulate the reality of life in home or workplace, where tasks are more complex and integrated and relationships more complicated? How, for example, can you easily teach thirty people in a classroom how to redecorate a room with any confidence that they could do it at the end of the lesson?

Why, come to that, do we make schools responsible for so much (the custody of our children, their education in almost everything, their certification, their behaviour and their training for work)? It never used to be so. One learned about behaviour, values and customs at home and about work at work, and one went to school to learn only those things, like reading, writing and arithmetic, that schools could do better. Maybe we should not blame the educational system. We have given it an impossible task – to be guardian, educator, trainer, counsellor and judge all at the same time.

The educational system has outlived the society which created it, a society in which an 'educated' minority managed a barely literate majority, who thought it enough to have a job and a living wage. Education, it is true, offered a bridge between the two groups, but it was a bridge that few took, even in later years. Halsey *et al.*'s research into family, class and education[3] revealed that in 1972 only 1.8 per cent of children from the homes of skilled, semi-skilled and unskilled workers (5.49 per cent of the population) entered university, compared with 20 per cent of the children of the professional class (13.7 per cent of the population). In the 1950s, they estimated, 40,000 boys each year who could have obtained 'A' levels were not staying at school long enough to do so. The danger is that things could get worse in a society where professional, managerial and entrepreneurial work is the growth occupation and skilled and semi-skilled work in the decline.

The diagram below was drawn by Barry Jones to describe the situation in Australia today.[4] The Australians inherited the tradition from Britain.

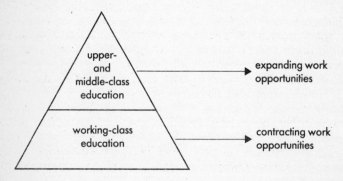

It was John Dewey, one of the fathers of modern education, who said 'The conception of education as a social process and function has no definite meaning until we define the kind of society we have in mind.' Society, this book argues, is changing. It is time for education to do likewise. Luckily, there are signs, few but definite, that it is.

THE SEEDS OF CHANGE

To many in the education business the charges in the previous section will seem not only unfair and oversimplified but also stale. The education business is masochistic enough to spend a lot of time picking at its own entrails, and the criticisms expressed above are not new. Many of them are aired in books like David Hargreaves's *Challenge for the Comprehensive Schools*[5] or John White's *The Aims of Education Restated*.[6] Many teachers will also claim that they have moved to correct many of the criticisms, although they are still trapped by the expectations of parents and by the examination system.

There are, indeed, many seeds of change around. Put them all together and we may be seeing a timely revolution in education, one which will answer many of the questions posed in the last section, among them the following.

Multiple criteria

The knowledge-based examination will soon be only one of many ways of certificating the young. The Department of Education and Science has proposed a 17+ qualification based on one-year courses for those staying on at school after 16 but for whom GCE 'A' and 'O' levels are inappropriate. Other examining bodies (such as the City and Guilds of London Institute and the Royal Society of Arts, Manufactures and Commerce) have been modifying the opportunities they provide in pre-vocational, technical and commercial qualifications. These initiatives could, however, backfire if they were seen as the hallmarks of the second-rate, to be looked for only if the traditional certificates were beyond one's reach.

Of critical importance, therefore, are two further initiatives. One is the move towards graded assessment, whereby proficiency in other subjects can be measured in the same way as music, with its graded examinations. Graded assessment, independent of age, is becoming increasingly common in modern languages. It needs to spread to other subjects. The other important initiative is the experiment with pupil portfolios. The idea is that every pupil should leave school with a portfolio of achievements. Part of this portfolio would be GCE examination results, if any were obtained, but also included would be any technical or commercial qualifications, graded assessments, reports on work experience and the 'profiles' that are used by some schools and authorities to build a more rounded picture of a pupil's strengths and personality.

No doubt for some time to come the most valuable coins in anyone's collection will remain those GCE examination grades, but the existence of other criteria, as a matter of routine, does allow employers, admission officers or tutors to ask for evidence on other aspects of learning and performance. Student portfolios (which are broader than 'profiles') are the outward and visible sign that education is not one-dimensional. They need every encouragement.

Education for capability

The mood of the age is an important element in shaping educational policy. Dr Arnold captured the mood of his age and class when he reminded his pupils: 'What we must look for here, is, first, religious and moral principle; secondly, gentlemanly conduct; thirdly, intellectual ability.' But Dr Arnold's precepts have lasted longer than they deserved in a changing world. There is another mood around today which is finding expression in different ways.

The Royal Society of Arts put it most forcibly in its public manifesto, signed by over a hundred of the great and the good in British society. It argued that comprehension and cultivation (Arnold's priorities) were receiving too much attention at the expense of areas like creativity, competence, capability and the ability to relate to others, 'communion'. Auriol Stevens, writing about comprehensive schools, puts it like this:

> It is of crucial importance that as many young people as possible become productive, as few as possible become dependent. It is not a vast number more academics – or even a vast number more Nobel prize-winners, excellent as they are for national morale – that we need. We need versatile, practical people, capable of managing their lives, contributing to the lives of others, earning their living and enjoying their leisure. The answer is not, therefore, to unpick the comprehensive system or even substantially to modify the principle, reverting to an older, academic approach . . . Between 20 to 25 per cent educated to a high level is no longer anything like enough, and academic skills alone are too narrow a basis on which to build a sound industrial nation.[7]

Other people take the same line. The Further Education Unit of the Department of Education put out a document, *Experience, Reflection, Learning*, describing an approach in which experience comes first, followed by theory;[8] the Schools Council document *The Practical Curriculum* suggests a framework for implementing it;[9] the TVEI programme of the Manpower Services Commission now finances pilot schemes for 14–18-year-olds in sixty Local Education Authorities, schemes which are designed to improve the technical and vocational education of pupils at all levels of academic ability.

Work experience

The charge that you cannot learn everything in a classroom is being met by the provision of more and more opportunities for pupils to experience real work. The Youth Training Scheme, introduced by the Manpower Service Committee in 1983, is a pathfinder in many respects. It is a specifically educational initiative (not a job-providing one) but bases itself on the workplace, with only a quarter of the time required to be spent in formal schooling, and it pays the student a weekly wage (admittedly a very small one). Whatever its teething problems, the scheme, in these two ways, turns part of our educational thinking on its head. You learn to work at work, it says, and learning is work that deserves to be paid for.

Actually, these two principles have been around for a long time. Apprenticeship, professional training (articles), the sandwich courses of vocational degrees at polytechnics and technical colleges all applied them, but at the unfashionable, practical end of the educational spectrum. Since the Second World War the system of student grants for higher and further education has extended the idea of a salary for learning.

That work is learning and learning work is an idea which is spreading. More and more educational authorities are recommending periods of work experience as part of the 14–18 curriculum. The concept of a student wage after 16 is being promoted by at least one political party. Community service as a form of work plus learning is advocated by many and practised by a few schools, although to make it compulsory, as some would, would turn an opportunity into a duty and a chore.

Regrouping

The fall in student numbers is forcing educational authorities to rethink the arrangement of their schools. No longer is it possible to maintain as many big, all-through schools with the full range of options at sixth-form level. We are beginning to see more sixth-form colleges, more sixth-form consortia between groups of schools and more ad hoc alliances between schools for certain subjects. Community schools are increasing in number, bringing adults into the classrooms as students, often alongside their children.

This regrouping adds variety to the system; it tends to make the units smaller and more different one from the other. Why should all schools be the same? An egalitarian approach should not have to mean a uniform approach, every school a mirror-image of every other, for no institution can be all things to all people. It was always a fallacy to imagine that the comprehensive principle required comprehensive institutions. What is important is that everyone has equal choice, but that can best be achieved by providing access to a range of institutions that choose to differ in their emphasis, their field of interest and range of topics, rather than only in their quality. If all schools are designed to be the same, they can be compared only in terms of quality, leading to a pecking order on a single scale.

Excellence and achievement need more than one face if we are to produce a generation capable of living portfolio lives. The regrouping of institutions should encourage each to concentrate on what it does best, giving more scope to the individual to choose in which area, and on what dimension, he or she wants to succeed.

The proliferation of seeds is encouraging. The danger is that they may be all planted in one seed bed, the seed bed of the 'academically less able', the seed bed of the second-class.

It may be argued that this is socially fair, that non-intellectuals should be given the education most relevant to the new age, as some sort of compensating mechanism. It can indeed be exciting to see new experience-based courses being so imaginatively organized in many comprehensives for the non-examination classes, and to watch the enthusiasm of the participants while their more academic colleagues sit in rows in front of a blackboard. But can it profit us or them to divide our youth in that way? Will it not still be a continuation of Plato's world in which the educated elite lord it over the workers? Power and influence will still go to the bright but not necessarily to the best. And do not the clever need to be capable too?

If we are to turn our frequently disabling system into a system that enables everyone to get the most out of a world where all will have to look after and look out for themselves to a much greater extent than in the past, where learning and relearning will have to be a continuing part of life, where saleable skills, the ability to deal with people and the capacity to express oneself are all crucial, then, first, capability needs to be part of everyone's curriculum and, secondly, respectable achievement has to come in more than one guise. Nobel prize-winners, by all means, but not to the exclusion of electronic engineers, designers, photographers, builders and, yes, teachers.

We have an opportunity presented to us by the current crisis. Work for most young people is a party postponed, often until they are 19 or over. We can use the enforced postponement to make a reality of the guarantee enshrined in the Butler Education Act of 1944, that of education to 19 for all who wanted it. It was a guarantee taken up by a shamefully small proportion of Britain's population compared with those of other countries (see box 6.2) because education meant the kind that got you into university.

It need not mean that; it should not. Universal, voluntary and paid education for all up to 19 would allow us to develop rich menus

BOX 6.2   15–19: A TIME FOR EDUCATION?

In 1980 the OECD provided a comparison of percentages of 15–19-year-olds enrolled in any form of education in twenty-three countries.

	Country	Enrolment (%)
1	United States	73.7
2	Japan	70.9
3	Switzerland	70.1
4	Canada	64.9
5	Norway	63.6
6	Netherlands	62.7
7	Belgium	61.3
8	Finland	60.8
9	Denmark	57.4
10	Sweden	56.3
11	France	54.6
12	Ireland	50.0
13	Greece	45.4
14	Australia	45.0
15	New Zealand	44.8
16	United Kingdom	44.6
17	Italy	43.9
18	West Germany	41.5
19	Spain	35.1
20	Luxembourg	33.5
21	Portugal	33.4
22	Austria	32.0
23	Turkey	12.7

The West German and Austrian figures are misleading because they do not include the vocational courses which 90 per cent of those leaving school attend as part of their apprenticeship. Only 6 per cent of West German youth enter a job with no training at all.

of alternative curricula, menus which would each have their own validity and respectability. It would cost money, but would it cost more than supporting the same people on the dole and having to keep many of them there for many more years because we would not have enabled them to live without it? Do we really have any option, given the job scenario of chapter 2?

A varied set of menus beyond 16 would do much to break the universities' historical stranglehold on the educational system. Higher education would still be crucially important, but it would not be the only passport to respectable work: 55 per cent of our population would no longer be educationally disenfranchised.

Education for all to 19, no matter how varied and respectable, would not solve the problem. Higher and further education cannot be immune to the challenge from the future. They too have questions to consider:

(1) Should they not look for breadth rather than depth in their incoming students? If specialization in the schools could be postponed, it would be much easier to create a broader and more varied curriculum, mixing work with learning for everyone.

(2) If universities and polytechnics complain that they would need longer to bring students up to the required levels (a questionable assumption in all those courses not taught at school level), should they then not look anew at the medieval arrangement of those short terms, which are today an unjustifiable under-utilization of physical and human assets.

(3) Is university education best sited before one's working life commences, or would many courses be better placed after a few years of work experience (as the graduate business schools commonly require)? Does the idea of educational credits – an entitlement to education to be cashed by the individual at a time of his or her choice – have anything to commend it?

(4) Should they not become less monopolistic in their degrees and allow more mobility through a system of transferable credits, as in Germany, which would also allow people to move in and out of higher education?

These are not new questions, but they have a new urgency today: working life is soon going to become more episodic for many, rather than a continuous process; retraining and re-education will soon become more and more important, and breadth of learning as important as depth. The seeds of change are, perforce, being planted where the 16–19-year-olds are, but the shoots produced by those seeds are going to reach out and up into higher education.

THE NEW SCHOOLS

Learning never did end with school, college or university, nor was it ever confined to the classroom or the lecture hall. These were only the convenient assumptions made by the formal part of the system. They allowed us to draw a boundary around education, the better to control it and monitor it. Like much else in education, as we have seen, this had its unintended consequence: if you did not get your education within those boundaries, you were effectively uneducated in the eyes of the world. In spite of the endeavours of the comprehensive school movement, effective formal education existed only for the young and for the minority.

One of the good features of the new world of work is that these boundaries around education are being broken down. New technology, new patterns of work, new demands for training and learning will break the straitjacket on education, create new kinds of schoolrooms, new breeds of teacher and new forms of assessment. We have already looked at some of the seeds of change in the previous section on schools. Let us consider some more.

The new students

Eighteen years ago I was charged with launching a one-year full-time programme of education in management for executives in their thirties. It was, in Britain at that time, an outrageous novelty. Only university professors took sabbaticals, and even these were not times for learning or for courses but for polishing off that book or that piece of research. Adults in the prime of life and of work did not go off to school, only to the occasional conference. Today serious study for adults is commonplace. In-service training for professionals, management courses for executives in all sectors of society, Open University degrees – these are now familiar. Universities and polytechnics welcome a proportion of 'mature' students on their major courses, no longer sending them round the corner to the Extra-Mural Studies Department (literally 'outside the wall' of formal education!). Education after school is big business, as box 6.3 shows.

BOX 6.3   DO ADULTS WANT TO LEARN?

The Advisory Council for Adult and Continuing Education conducted a survey of 2,500 adults between 17 and 75 years old in England and Wales in 1982. These were some of their findings:

More than half of the adult population was seriously interested in taking courses. One fifth were already on courses; a further third would probably do so if they had the opportunity.

Six out of ten men named work as the reason for taking the course. Women mentioned challenge, meeting people and general interest.

Forty-six per cent said that they wished they had done further study in the past but were put off by lack of money (one-third), lack of knowledge (one-fifth) or lack of qualifications.

Middle-class people tended to value education and self-improvement more than working-class people, who preferred to study at home.

Fifty-nine per cent felt that post-school education should be subsidized, but 47 per cent were prepared to pay part of the cost themselves, and 39 per cent thought that people should pay the full cost.

Eighty-eight per cent were in favour of a formal entitlement to adult education, but only 50 per cent would be likely to take it up.

Adults: Their Educational Experience and Needs
(Advisory Council for Adult and Continuing Education, 1983)

The home as classroom

The Open University brought school into the home, even into the bedroom, with its late-night and early-morning programmes. The personal computer and the video recorder, as they get cheaper and as commonplace as the television set or tape recorder, will increase the possiblities for distance learning for students of all ages. Homework becomes both more interesting and more substantial when films, computers and tutorial tapes are there to help you. Students can even be tutors for each other in many instances. Management courses depend on the learning syndicate. The informal student study groups of the Open University are an important complement to the formal tutorial system. Much of the voluntary world depends on self-help and mutual aid, whether it be Alcoholics Anonymous,

Gingerbread (the single-parents' association) or the tutorial groups of marriage guidance. Parents coach their own children, or, more easily, other people's children.

Take the mechanics of learning out of the classroom and you begin to chip away at the professional monopoly of teaching. Not everyone can design a course of learning, create exercises or give lectures and demonstrate – all those must remain professional skills – but almost everyone can help someone else to learn (often better) because he or she has experienced or is experiencing the same problems or difficulties. Mature citizens can be mature tutors as well as mature students, and there are going to be a lot of mature citizens with time on their hands. Why cannot some of them be part-time tutors for our kids?

Moving school into the home may at first seem threatening to the professional teacher, but it can be a way of liberating time and space. If more work were done at home or in groups of homes, super-vised/tutored by non-professional adults, schools and colleges could devote more time to the design of material and the provision of feedback, rather than to transmitting information. Learning prog-rammes could be more readily tailored to the individual than is possible in a class of thirty or a lecture hall of 100. Schools could be used, physically, twice over – in the mornings for the young, in the afternoons and evenings for the more mature (rather as community colleges are today).

Whether or not the schools and universities take advantage of it, the home is the sociological university of the majority of the population. It is through the television set and magazines that most of us form our impressions of what goes on in the world, of other people's values and priorities, of what is generally accepted as right and what is not. We may not take examinations in what we learn, but we are all today the beneficiaries of television's mandate to inform and educate at the same time as it entertains.

The workplace as school

It was always true that you learned the skills required for work *at* work, whether it was called apprenticeship, articles, traineeship or probation. Only recently has there been a shift away from this, with more demands being placed on schools and colleges to prepare people for work with more vocational courses and more educational experiences geared to the world of industry, commerce and the work organization. With hindsight, expectations were unrealistic. Schools and courses can transfer information; they can teach particular skills and drills; but they cannot provide the 'coached experience' which is the crucial element in all forms of apprenticeship or personal development at work.

Long years ago, Professor Katz, in the *Harvard Business Review*,[10] distinguished between three types of skill: technical skills, human skills and conceptual skills. All three are necessary at work and in life, but whereas technical skills can be 'taught', human skills can be learned only by experience and conceptual skills only developed. Classrooms, therefore, are best for technical skills, but experience and coaching are needed for the other two. Putting too much faith in classrooms and in what can be taught runs the risk of putting too much emphasis on technical skills and too little on the others. Life cannot be reduced to a textbook.

The Youth Training Scheme implicitly recognizes this in reverting to the traditional mix of work experience and formal training. Increasingly we may expect the workplace to resume its role as the school for work, because it is in the end, with all its possible faults, the best school for work. It is also the cheapest, a fact which will not have escaped the notice of Governments, and it is a school that is already staffed.

The new development is likely to be the emergence of the workplace as the school for life *beyond* employment as well as the school for life *in* employment. As the effective retirement age comes down, and more and more people look forward to a working life that extends beyond employment, they will be seeking to equip themselves with useful and saleable skills. Organizations already run preretirement courses to help people to adjust to life outside the organization, but these are too short and too skimpy to meet the new needs for new skills. The Army used to advertise 'Join the Army and Learn a Trade', implying that you could learn a skill which would be useful not only in the Army but outside it when you left. It is the kind of attitude which more and more organizations should start to promote. If they do not do it voluntarily, it may be forced upon them, again because they are likely to be the cheapest vocational schools around. A mandatory part of any redundancy or early retirement scheme could be a reskilling course, part-subsidized by the state.

In any case, as organizations become more professionalized they will, in their own interests, need to update their various professionals and skilled people so that more and more people will leave organizations fully 'credentialled', equipped with a trade, rather as an accountant with a large firm of auditors can leave and practise his profession on his own without any further training. One way or another schooling is going to be a larger part of an organization's responsibility.

But organizations may find certain things beyond their scope. In particular they are unlikely to be the best schools for self-employment or entrepreneurship. The technical skills they may be able to teach, but not the attitudes or the ways of behaving that one

BOX 6.4   NEW SCHOOLS FOR ENTERPRISE

Among the new schools of Britain are the growing corps of projects and courses to help the unemployed to become self-employed. Schemes described in Richard Bourne's and Jessica Gould's *Self-Sufficiency 16–25* include the following:

The New Enterprise Programmes – four courses funded by the Manpower Services Commission at the business schools in Manchester, London, Durham and Glasgow to train would-be entrepreneurs in the basic skills of business and to help them launch their first idea.

Project Fullemploy, Clerkenwell – based in the small-business world of the Clerkenwell workshops, this course is directed specifically at people aged between 17 and 24 with low academic qualifications and is designed to help them become self-employed by training and counselling.

URBED – a training centre for those wishing to start their own businesses and a small firms support network once they have got going.

Business Opportunities Programme – financed by the Inner London Education Authority at the City and East London College for which applicants need no prior qualifications.

Hartlepool Co-Operative Enterprise Centre – the centre pro-vides subsidized space for would-be businessmen but also gives advice and acts as the core business, buying the tools and paying the wages in return for profits from the new businesses.

New World Business Consultancy – set up at Camberwell Green to help the black community form new businesses.

Bootstrap – a project based in Hackney to start small co-operative businesses with unemployed volunteers, who receive help and advice but no wages until the projects are viable commercially.

Job Creation Limited – this firm guarantees to create new jobs for firms or local authorities and is content to be paid by results. It concentrates on providing facilities for the self-employed, together with proper market research to fit the individual's skills and ideas to a possible market.

Richard Bourne and Jessica Gould
Self-Sufficiency 16–25 (Kogan Page, London, 1983)

needs to initiate enterprise. Organizations, by and large, are responsive creatures, reacting to circumstances and events. The person who would found a new business, initiate a new venture in the community or launch himself or herself into self-employment needs to create, not respond. Risk and uncertainty become a necessary part of life; networks, contracts and telephones are essentials. It is one thing to work your way through an in-tray, quite another to sit down in front of a blank sheet of paper. The rules, procedures, committees and systems of organizations may have frustrated you for thirty years, but living without them can immobilize you. The large organization is not always going to be the best school for enterprise, but there are alternatives, some of which are listed in box 6.4.

The new credits

Education after school is very conventionally organized if it is a degree that you want. Generally you join an institution for two, three or four years – an institution which sets its own rules of admission and its own standards of graduation, does not encourage intermittent study (one year on, one year off) and does not recognize any study you might have done elsewhere. This is not universally true, however. The Open University does accept intermittent study and does make allowances for courses taken elsewhere. There is also a system of external assessment which is intended to make sure that degrees in similar subjects in different institutions maintain comparable standards.

The Open University is a notable exception. As the most recent addition to the institutions of higher and further education, it has tried hardest to adapt to the needs of the individual in today's world and to the realities of life. It has realized that if degrees are restricted to people who can give up three or more years full-time and continuously to one institution, this form of education is effectively confined to people who have not started work. Later on the logistics for most people get too complicated.

The new patterns of work suggest that more people will need formal further education at more periods in their lives. But if they are going to be able to take advantage of this opportunity, then the institutions will have to become more flexible. The idea of accumulated credits as the basis for a degree, which is common in the USA, in Germany and in some other European countries, should be accepted in Britain. A credit system would allow individuals to study courses at a range of comparable institutions and over an extended period of time.

Credits do not have to be confined to academic course work. Norman Evans of the Policy Studies Institute has argued for a system of experiential credits in which recognition is given for experience in different forms of work.[11] This has been the practice for a number of years in the community colleges of the USA. These colleges award two-year associate degrees, which then count as equivalent to the first half of a four-year full degree if anyone wants to proceed that far.

The first steps towards experiential credits happen informally. More and more institutions are prepared to relax their academic entrance requirements for students with appropriate experience. In certain areas (e.g. management) some appropriate experience is almost a prerequisite. It is more common today for students to postpone their university studies for a 'gap year', often with the positive encouragement of the admissions tutor. It is not a big step from that to the award of formal 'entry credits' for relevant experience, community work or travel in addition to academic grades. Naturally, if gift work and pocket-money work were seen as educationally valid and useful, it would help to take some more of the pressure off youth unemployment. Experiential credits for the young would be one important way to fill some of that yawning gap between 16 and 24 which will confront the young people of the next twenty years.

Experiential credits are also important for the older student. Not only could they shorten the total period of study required for a qualification, if recognized as formally equivalent to course credits, but they would also do a lot for students' *amour propre* or self-respect. To enrol as a student today is to leave behind you, when you enter the lecture room, any authority or experience which you have acquired through work or pastime. It is to start afresh, as if one were 18 again. This kind of ego-stripping may appeal to the masochists among us or to those who want to escape from their past, but to many it is an unnecessarily high and cruel price to pay for education.

There are problems, of course. How does one evaluate the quality of experience as opposed to its quantity? Some may have worked in a bank or an office for twenty years and learned nothing. It is possible to travel without opening one's eyes, or to care for the weak without opening one's heart. The community colleges of the USA rely largely on interviews or on reports from supervisors at work. Maybe these subjective measures are no more inaccurate than examinations sat in draughty halls, which depend on memory for information and a pen for communicating it.

Last, but never least, is the question of finance. The idea of financial credits for education has been around for some time[12] and has been supported as an idea by the Commission of the EEC. The idea is that at the age of 18 everyone is given an index-linked entitlement to money as a grant towards three years of further education, to be cashed in at any time in his or her lifetime. This would make a reality of the commitment in the Robbins Report on Higher Education to provide higher education for all who wanted it and were qualified to obtain it. Robbins said nothing about age, but it is that implied age criterion (after school and before work) that has excluded so many.

What would be the effect of such an age-free entitlement? How many would postpone their education? How many would, in fact, take it up in later life? Could it be afforded? Should it be means-tested, or would that push people back into opting for education while they were young and still poor?

We cannot tell, and for that reason a full system of financial credits seems unlikely. It would be an open-ended cheque. But the principle behind the idea can be accepted and implemented more informally, more piecemeal, more gradually: that is that everyone, of whatever age, should be able to receive formal education after school if he or she wants it badly enough. For that to happen we need to loosen up the whole system; we need to broaden the criteria by which we measure learning, allow people to learn at their own chosen time and pace, increase the flexibility to move within and between institutions and build on all the technological possibilities which allow people to study away from institutions for part if not all of the time. It will be inconvenient for the institutions. They will not do it unless they are pushed. Therefore they must be pushed, by government pressure, by popular opinion or by economics and the need to find more clients.

THE NEXT STEPS

This chapter has been a plea for more flexibility, more variety, more choice and more participants in education. Future patterns of work will simultaneously demand that more people learn more and provide them with the opportunities to do it. Will the supply be ready for the demand? Not if schooling remains such a law unto itself, such a closed monopoly, shaped by its past. We have noted the kinds of things that need to happen if education is indeed to be the growth sector of the immediate future and the infrastructure for our future wealth (of all types).

Behind these new initiatives lie certain common principles. It is, in the end, more important that we agree on the principles than on the particular initiatives because there must be more ways than one to achieve our desired ends. Let us therefore set out the principles, so that they do not get lost in the detail.

(1)　More people will need and want more education at more points in their lives. For economic, social, aesthetic and moral reasons we ought to do all we can to meet their needs. Education is an investment, not a cost.

(2)　Learning has many dimensions, and success must be measured on many scales. One scale should not be more important than all the rest. We need a greater variety of assessment and broader criteria of success at all levels . Capability, creativity and community are at least as important as culture and comprehension.

(3)　Learning does not happen in one place only, and that place a school. The home, the workplace and different types or varieties of school all need to be blended together in an educational network.

(4)　Individuals should have more scope to arrange their own network by choosing from a greater variety of schools or universities through systems of transferable credits or within consortia of schools. Comprehensive opportunity does not necessarily entail comprehensive organizations.

(5)　The work organization must recognize its role as the principal school for work, both during employment and after it.

An education system is a mirror of a society. It is unrealistic to blame it for the flaws in that society or to expect it to be the lever for change, although one or two brave spirits may experiment at the edges. It is no part of the argument here that education should lead the way into the future, only that it adjust its mirror quickly enough to reflect the world that is springing up around us rather than the world that used to be. All the opportunities of the future of work depend upon a population with access to education, a people geared to think and act for themselves, a nation that will rejoice in choice and responsibility, not shun them. It we do not get education right, quickly, we will be faced with a scenario of lost opportunities and a generation of whom it might be said one day, 'They have a bright future behind them.'

7

Choices

I have argued throughout this book that the future will be different, that it may be dangerous but also exciting and life-enhancing. It is not predetermined. We are not in the grip of some technological monster or some invisible hand of economics which will force us down a certain route. On the other hand, we do face constraints. We cannot ignore technology or economics, and we should have accepted by now that the status quo cannot be the way forward – we shall have to change one way or another. How we change is, however, largely up to us.

In this chapter the emphasis is on the individual. What room will he or she have to twist and turn in tomorrow's helter-skelter world? Must we, as individuals, leave it to governments, the captains of commerce and the media messiahs, or do we have choice? The answer has to be that, paradoxically, in a state of flux there is more room for individual choice than ever before, *but* that in such a state of flux people often feel incapable of choosing. Perhaps if we could focus on the options more clearly, choice would be easier. This chapter is an attempt to provide that focus, looking in turn at the questions that every individual could and should be thinking about:

What sort of life do I want?
What sort of family and home do I want?
What sort of society do I want to live in?

Whether we, collectively as a society, get the most out of the future of work will depend very largely on the way in which each individual answers these essentially personal questions.

WHAT SORT OF LIFE?

'Life planning' is a set of exercises designed to help you to work out where you want to go in life and how to get there. The opening exercise is often an instruction to draw a line representing your life with all its ups and downs and to continue it on into the future.

you are here

Try it some time.

It is interesting and instructive to find out what have been the particular highlights and depressions of our lives. Often they will have had nothing to do with conventional success or material prosperity. It is intriguing to guess where the line might lead in the future and how long it will last. For all but a very few the line of the future always goes upwards in our own minds. We are at heart, it seems, incurable optimists, however much we may lament the future in public. Most people too see the past as having been done to them but feel they have much more influence over the future – but then that is the point of the exercise, to help to give people more control over their destiny. Young people (under 25) often dislike the implications behind the exercise: 'I don't want to decide what I'm going to be or do until I've seen what happens,' they say.

What kind of lifeline may we expect in the future, and what will be the meanings of the peaks and the troughs?

The changing shape of work and the erosion to the employment culture in which we grew up will affect our family life, our assessment of the meaning of work and, ultimately, the kind of life we want to live. It will not be all the fault of work, of course. Other things are changing too: our set of values, our concept of the individual and particularly of the role of women, our expectations of the state and its responsibilities, even our notion of our international involvements and obligations. The changing structure of work acts, however, like a lens on a camera, pulling all these elements into focus in one picture.

The individual growing up today and leaving school in the next few years has a future to look forward to that is very different from his or her parents'. This should be evident from earlier chapters, but it is worth underlining again the main points as they will hit the individual.

The 50,000-hour job

How people distribute their 50,000 hours will be a matter of choice and circumstances.

Some will *pace* their lives, working part-time (twenty-five hours a week) for fifty years, with eight or ten weeks' holiday each year.

There will be more such jobs, particularly in the service sector, but they will not be particularly well paid, nor of high status.

Some will *peak* their lives, working hard and long in the early years, from 25 to 45 maybe, then withdrawing to a second career and to a role of wisdom rather than energy. Professionals and managers will fit into this category, particularly if the age of responsibility were shifted downwards in institutions so that people moved into key positions in their mid-thirties.

Others will choose, or be forced, to adopt a more intermittent pattern, moving in and out of the formal workforce with time away for education, child rearing, home building or enforced leisure.

All will have to stretch the money which they earn in 50,000 hours over the rest of their lives, or try to.

Part-time work

Part-time work has long been the dirty, menial, unprotected side of work. It is generally low-status, badly paid and done only by those who can get nothing else or cannot (because of responsibilities to families, for instance) take on full-time jobs. That, at least, has been the public image, and much of it is true. Part-time work, often for a fee, has been the effective way of reducing the cost of a job in many businesses – that is one of the underlying themes of this book – but it may not all be bad news. The *General Household Survey* of 1980 spends considerable time analysing the unsuspected finding that part-time workers are the most satisfied of all workers. This is partly due, it concludes, to the fact that part-time workers are older (over 25) and do not experience the acute dislike of work which many young people express. In part it may be that women working part-time have lower expectations of the job. They go for the money, to get out of the house, for the companionship. Indeed, as Judith Humphries notes in her description of part-time work,[1] there is ample evidence that part-time workers are able to sustain tedious work well below their level of competence and demonstrate great good humour and cheerfulness in so doing.

Judith Humphries's own research suggests that many part-timers get considerable satisfaction out of their job and, indeed, work at it with more enthusiasm and energy than the full-timers. The problems are not so much the money as the design of their work and their standing in the organization. If they have their own clear areas of responsibility – this room to clean, these people to care for, this batch to process – satisfaction is higher. If they are not seen as marginal people, external to the system, but are included in the communications systems and the social networks, then the greater sense of freedom usually more than compensates for the fact that they cannot hold key jobs or managerial roles. It seems to be that if you go into part-time work because you want to, knowing the snags, then it can be a very satisfying part of life. There is, after all, good evidence that most people do not want to work for as long as they do, whatever they may say in response to surveys. Across northern Europe the average worker takes around twenty-three days in absenteeism or sick leave – almost 10 per cent of the full working year. Perhaps if it were properly organized and better paid, many more of us would like our work to be part-time.

The argument of this book is not only that there will be more part-time work in future but also that the work will often be of higher status and of more importance to the organization and therefore

BOX 7.1   WEEKLY TIME AT WORK AND LEISURE: A SCHEMATIC VIEW FOR A MAN WORKING FULL-TIME

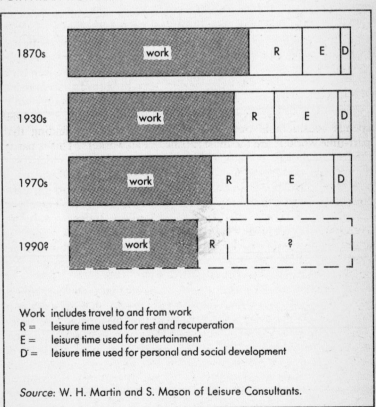

Work includes travel to and from work
R = leisure time used for rest and recuperation
E = leisure time used for entertainment
D = leisure time used for personal and social development

Source: W. H. Martin and S. Mason of Leisure Consultants.

better paid. In recognition of this, organizations will speak more of freelance arrangements, of contract work, of job-sharing or of shorter shifts. It will still be fees in place of wages, still part-time contracts, but the new words will distance it from the old image of the cleaning corps, the Saturday help and the refuse disposers that has dogged part-time work, just as 'working from home' sounds better than 'homeworking'.

The status and pay of part-time work may improve but it will still be less remunerative, less secure and less of a 'career' than full-time work. It will spread the work around more thinly, making some work available to more people, but it will be a positive option only for people with other things to do in the rest of their lives. Most obviously, these will continue to be people who have domestic responsibilities – but do these always have to be women? Increasingly we may see organizations employing their own retired (or about-to-retire) people part-time. It will be done as a way of getting the core of their work out of them for less cost, but it may suit the individual very well to trim down his or her job commitments at that stage in life, to allow room for other interests, without an abrupt leap into retirement.

Indeed, in the ideal society people would do part-time work not because they had to do it but because they did not have the time or the need to do more. That is true today only of some older people, of some parents, of some young people. It may be that the spread will increase. The story of the Irishman who was paid double by his new firm and therefore stayed at home on Thursday and Friday 'because now I don't need to work so long' might be true of more of us if we had the option.

Discretionary time

Fewer job hours must entail more discretionary time – as box 7.1 demonstrates graphically. It would be wrong to think of all this time as leisure. 'Leisure', in fact, is a misleading term because it can include rest *or* recreation *or* self-development. Interestingly, the Dutch do not have a single word for it. Perhaps we should not either. It is usually seen as the antithesis of work, so that if you do not have work, you do not have leisure either. Indeed, to speak of the unemployed as having ample leisure would be regarded by them as an insult, patronizing in the extreme. Imposed leisure is neither rest, nor recreation, nor development.

What will we do with all this discretionary time? Chapter 3 suggested a range of possibilities, most of which could and should be termed work. A study by Ann McGoldrick of 1,800 early-retired men and their wives[2] grouped the possibilities in an interesting way. She identified the following types:

(1) 'Rest and relaxers'. These people, mostly the older ones in the sample, were content with the traditional pastimes of newspapers, TV, walking, gardening and trips in the car ('It's nice . . . to be able to do as you please all day').

(2) 'Home and family men'. These enjoyed spending more time with their wives, looking after and helping in the home.

(3) 'Hobbyists'. For many the hobby had become the focus of their lives. Hobbies ranged from DIY, philately, birdwatching, golf and fishing to music and art ('I can now follow my hobbies properly. In my case its music. My days are so filled that I could not fit in the adult education course I intended to follow').

(4) 'Good timers'. Social life, travel and going out in the evenings were the choice of some. One foreman went on a six-month trip to New Zealand ('Travelling is one of our major interests and we've had two major holidays already this year').

(5) 'Committee and society men'. Twenty-four per cent of the sample found that they were able to devote more time to societies and committees that they had been interested in before ('I am chairman of two local bodies and treasurer of another, so I'm not idle for much of the time. They need my time').

(6) 'Volunteers'. Nineteen per cent of the sample had done voluntary work after ending employment. This ranged from helping friends and neighbours to full-time work for a local charity.

(7) 'Further education men'. Nine per cent enrolled on courses at universities, among them the Open University, and at local colleges ('I've gone back to school again. I'm balancing my engineering training with a degree in Humanities at the Open University').

(8) 'Part-time jobbers'. Twenty-four per cent did some extra work, a few because they needed the money, but most because of the interest it gave them. More intended to do some part-time work when their year of unemployment benefit ended ('A part-time job is an immense benefit – it's a hedge against inflation and keeps one active mentally').

(9) 'New jobbers'. This includes the 'easier jobbers' who took a lower and slower job for less pay, the 'other job men' who looked for equivalent jobs, the 'second-career men' who wanted the chance to move into another career, usually after a period of retraining, and the 'entrepreneurs', who used early retirement as a chance to start their own business with redundancy money as capital. In many cases they varied the time devoted to the job, in one instance closing the business down for the cricket season!

In general these early retirers described themselves as 'less tired' (48 per cent), 'more at ease' (70 per cent), 'in better general health' (52 per cent), 'under less stress and strain' (75 per cent), 'less worries' (68 per cent). But there was a minority of the disenchanted (particularly those who were sick), who had financial problems or had been forced into retirement.

This picture of people in early retirement is likely to become more general as more people move out of their career employment in their fifties or even their late forties. It is a picture of full portfolios, for the most part, although differently balanced from individual to individual. It is, however, clear that if the move is voluntary, planned for and done at the right age (not too old) and with the right money, then the readjustment from the job to a portfolio of activities works much better. The moral seems to be: don't drift – *choose*.

Achievement

In a society built around jobs, achievement is measured in terms of the job, officially at least. The *curriculum vitae* of the average individual is a list of jobs with a small space at the bottom for 'other interests'.

Get close to people, however, and the reality is often different. Ask men or women in middle age what they are proudest of in their lives and they are as likely to mention their children, their homes, their work in the community or their friendships as their jobs. Ask people in retirement and the gap between the *curriculum vitae* and the reality they feel is even more striking. People beyond employment see employment as a school for life and put it behind them, just as those in mid-life see school as a preliminary to work. If they talk about their schooldays too enthusiastically, you would wonder if they had achieved anything since school.

As employment becomes a smaller part of most people's lives, more people may find more ways of describing their achievement in life. Human beings are inventive rationalizers. We shall come to terms with the inevitable and turn it to good account. But it will take time. The redundant steelworker in his forties and the 20-year-old who has not yet found a job are entitled to laugh cynically at the suggestion that there is much more to life than a job, for it remains a necessary, although no longer a sufficient, condition of a successful life. In other words, life without the chance of a job is, for most people, unthinkable, but life that is all job will, increasingly, seem a confined sort of life as more and more people come up with more and more alternative definitions of achievement.

'I'm sure I've met you before,' I said to the man at a party. 'What do you do?'

'I'm a banker with — ,' he replied.

'That can't be it,' I said. 'I don't know anyone at —. What else do you do?'

He paused, blushed crimson and then said, apologetically, 'I don't do anything else. That's all I do.'

Ten years ago to be a banker would be enough. Today one apologizes for not having other interests and activities.

We can already see signs of creative rationalization in the Western world. Materialist values, which are closely related to job achievement, are no longer quite as dominant. To be self-sufficient, to live ecologically, to care for peace and justice – these are some of the boasts of the new generation, and who is to say that they are wrong? Many writers have pointed to a general shift in society's values. Gurth Higgin, for instance, suggests that the 'social project' of society is changing.[3] In medieval times religion was the name of the game and the Church the dominant institution. For the last two hundred years it has been economics, with the industrial organization pre-eminent. In the future it might be human growth, suggests James Robertson, or a SHE Society (see box 7.2). Fred Hirsch has argued that materialist societies can, in effect, run out of demand.[4] Once we have all that we *need* we compete for what we *want* until in the end the only things worth really striving for are 'positional goods', things which set us apart from others, like privileged education, a house with open views, a club which others cannot join. But, by definition, positional goods are rationed. If everyone could join the club, it would no longer be worth joining. Therefore people either get vicious or they give up the race and settle for other delights – like free time, personal growth, expressions of creativity, socializing or sport – which are unrationed and non-competitive, in that we can all have lots of each.

Ours should be a 'pluralist' society, then, with plenty of scope for different definitions of achievement and of the point of life. That will not necessarily make it easier for anyone, particularly in the early stages of life. A menu of possibilities looks more tempting than a set meal, but how does one choose if one has not sampled all the dishes beforehand?

BOX 7.2   THE SHE ECONOMY

James Robertson, in his book *The Sane Alternative*, extols the virtues of the SHE economy (Sane, Humane, Ecological). A SHE economy would put people before things, recognizing that people's energies and skills are important renewable resources, as contrasted with the unrenewable resources needed for capital-intensive plants. A SHE economy world would reverse the economic equation. In traditional economies employing people is a cost which organizations try to reduce, whereas in a SHE economy the cost would be the loss of opportunities for personally satisfying occupations. The emphasis would shift towards meeting people's need for belongingness and love, for esteem and self-actualization.

In a SHE economy people, suggests Robertson, will be less willing to work at jobs which they perceive to be personally frustrating, socially or ecologically damaging or futile; they will insist on spending their working time in ways that contribute to social well-being and their own personal goals and values.

A SHE economy will aim for healthy rather than cancerous growth. It will not be growth at any price. Human needs, social justice and ecological sustainability will be the priorities, not material gain. It will aim at what Schumacher called the 'economics of permanence'.

Coping with flexilives

Multiple possibilities is an optimistic way of describing uncertainty. A flexilife will not look like heaven on earth to everyone. It is, in fact, the kind of existence that the last few generations of women have been well used to, moving between work and family, mixing part-time work with home responsibilities, balancing career priorities with a concern for relationships in the home and, in many cases, having to abandon one for the other. It has been one of the arguments of the feminist movement that this is unfair, and there have been many who have chosen to escape from the flexilives of women into the career lives of men.

It is ironic that just as women have begun to win their fight to lead the kinds of lives that men lead, those lives are beginning to shift towards the pattern from which women are escaping. Some feminist leaders, however, are now beginning to say that the real challenge for women is to reconcile their new freedom with their need for love, family, children and a home. All of us, says Betty Friedan,[5] both women and men, must recognize the centrality of the family in our lives. That means breaking the general stereotypes so that all of us share in the dilemmas that women alone have faced until now. There is, claims Betty Friedan, some small evidence from the generation of teenagers in America that stereotyping is beginning to break down.

If, however, we are going to be the choosers and not the victims of the multiple-option flexilife, we shall probably need three things: a sense of person, a sense of purpose and a sense of pattern (the three Ps).

First, *a sense of person*. It is essential that each of us sees himself or herself as someone with an identity, someone who matters, who makes a difference in some way to some people, who can contribute and can create in ways however small.

The only psychological theory to have had demonstrable results in changing behaviour is reinforcement theory and its offshoots. Reinforcement theory got itself a bad name when Dr Skinner turned it into a synonym for manipulation. It does not have to be that. Essentially, it suggests that anything that builds up our self-image is liked by us; it makes each of us more of a person, and we will therefore repeat the behaviour which earned that reward. We need, in other words, to think well of ourselves, and on the whole when we think well of ourselves we perform better. Several studies have shown that high expectations lead to high performance in the classroom.[6] We all like to think we are OK. In a recent study of American males all of them, 100 per cent, put themselves in the top half of the population for 'getting on well with other people', while 25 per cent put themselves in the top 1 per cent.[7] Are we so very different from them in our self-estimations?

But our egos are fragile and easily bruised. No reinforcement can in the end mean no sense of person. Unfortunately, the prevailing culture in many institutions and in many homes sees it as better to deal out reprimands than praise. Reinforcement, or encouragement, is regarded as flattery, somehow rather wet. Britain's culture can be a bruising culture to many, particularly to that 45 per cent who leave school in mid-adolescence with no qualification worth the name, no job and no chance of one for several years. Too many have no sense of being a person. Too many, therefore, see choice as a curse, not a blessing.

Second, *a sense of purpose*. To choose between alternatives one must have some criterion, some view of where one is going, otherwise the choice becomes mere lottery, the toss of a coin or the easy way out. Materialism at least gave everyone a measurable goal of sorts. It may not have turned out to be all that wonderful when you got there, but at least travelling had a point. What the psychologists call 'traction', being pulled along by something, seems to be a necessary condition of psychological well-being, irrespective of what that something is.

Lord Ritchie Calder used to tell the story of the sun-tanned youth lying on the beach at Naples who was confronted by the American businessman. 'Why do you just lie there doing nothing when you could be fishing? That way you could sell the fish and buy yourself a boat and then in time a proper trawler which might become an ocean-going fleet until you were as rich as Onassis.' 'And what then?' asked the youth. 'You could spend the day on the beach at Naples.'[8] The story is intended to point to the futility of ambition, but at another level it illustrates most people's need for a goal. How long, one may ask, would the youth be content to lie on his beach without some purpose in so doing? Naturally, there are many purposes in life other than materialism. The desire to create, whether in art or engineering, the will to care for others, to live a life of dedication to God in a convent or a monastery, to learn and to educate oneself – these are all purposes which provide traction and the criteria for choice. To be purposeless is to be a potential victim, not a chooser.

Third, *a sense of pattern*. Choice involves change. Change is the more tolerable the more it is under our control and part of a predictable pattern, so that we can live with it even if we do not like it. Patterns come in various shapes and sizes. There is the pattern of predictability, as with the seasons. The onset of winter is bearable, if not likeable, because we know it is coming and it will end. There is the pattern of ritual, which suggests that an event is in the natural way of things which happen to all of us and is a proper staging-post in life. Marriages and funerals are important rituals for this reason, as is the retirement party in the works or the office. More important, however, are the patterns of relationships which provide the base from which we explore life, which are our insurance against disaster. The family at its best, the marriage, long-standing friends, neighbours in the community, the people at work – these are, or have been, the people patterns in our lives.

Today all the patterns are changing at once. Careers, like marriages, are more prone to abrupt endings. Work is less predictable and so, therefore, are earnings. The new staging-posts in life – leaving home, divorce, redundancy – do not yet have their rituals and therefore seem to each person something that is happening to him or her alone, not part of a pattern. Even families and other relationships are more fluid, less of a safety net, than they ever were.

In conclusion, take away from any of us our identity or sense of person, make our purposes in life seem futile or unobtainable, disrupt our relationships and the predictable patterns and rythms of our life, and we collapse. It was the method of the concentration camp and is part of the tried and tested preliminary to brainwashing. To destroy the three Ps is to turn human beings into victims, powerless and vulnerable. Such people are incapable of choice, particularly the sorts of long-term, decisive choice listed in this chapter.

Consider, then, the unemployed. Studies show (see box 7.3) that

BOX 7.3   THE SCOURGE OF UNEMPLOYMENT

Peter Warr of the University of Sheffield has listed the nine likely psychological effects of unemployment, as established by research.[9]

(1) Financial anxiety – two-thirds of working-class people have their income reduced by a half or more.

(2) Loss of variety – there are less places to go to, less things to do. Most unemployed people sleep more and do more housework and television watching.

(3) Loss of traction – there is less structure in life to draw one along, fewer goals and tasks that have to be done.

(4) Reduced scope for decisions – there is more freedom but less to be free about, fewer options.

(5) Less skill development – there is usually less outlet for one's skills.

(6) More psychological threats – more rejections from job interviews, credit applications and social meetings.

(7) More insecurity – particularly about the future.

(8) Less interpersonal contact – social contacts are cut because of lack of money and there is no work to go to.

(9) Loss of status – and the self-concept that goes with a work role.

These factors, Peter Warr points out, do not apply to all people equally. Those who were less committed to the job, who were older or younger, those who were middle-class, female and healthy, these all seemed to suffer less than the others. To be working-class, middle-aged and totally committed to your job was bad news when unemployment hit.

everything combines to threaten their identity, their perceived purposes in life and their accustomed patterns. No wonder that they are quiet and acquiescent; no wonder that they feel helpless and powerless, unexcited by calls to see work as more than a job, to think of unemployment as extra leisure time or to become self-employed entrepreneurs. Cynicism and a bitter laugh are the understandable weapons of the powerless.

The important thing, then, is to restore to those people their three Ps or, better still, to make them so resilient that they do not lose all when they lose their job. Portfolios can lose certain of their contents and still survive. To give the unemployed a job of sorts would help in the short run but would be a plaster on the boil, not a cure.

It could be argued that the three Ps are middle-class attributes, in that the middle classes have been reared to think of themselves as competent, to plan ahead, to sacrifice the short term for the long (which they would call investment) and to cultivate a variety of friendships and family networks. To the extent to which this is true, the middle classes will do better in the new world of choices, but more truthfully it is those who *think* as the middle classes are supposed to think who will more easily survive and thrive. It has therefore to be the aim of our families, our schools and our social institutions to give everyone a sense of the three Ps as far as that lies within their power.

That will not be easy. Schools, for instance, will need to change not only their curriculum and their culture but also their whole assessment philosophy from that of the racecourse, where only the first few runners count, to that of the mass marathon, where everyone who finishes wins. Organizations will need to help individuals to plan ahead, to take more responsibility for planning their futures, training for those futures and financing them, and will need to produce realistic and truthful contracts of employment as a proper basis for that planning. Families, finally, will have to resume more of the responsibility for managing the beginning and the end of life, but for that to happen society has to face up to the changing shape of the family, as the next section emphasizes.

WHAT SORT OF FAMILY AND HOME?

If the job becomes less important to many of us, so will the community provided by the employment organization. To many people that will be a big loss, for work to them will have been their major source of friends and of patterns of relationships. Since few want to live as social isolates, we are going to cast around for other communities to which we can belong.

For some these new communities will still be built around work, be that a part-time job or perhaps marginal or gift work in the informal economy. The new co-operatives that we must hope will spring up, the networks of the self-employed, the voluntary organizations, the arts and sports clubs, these will all provide alternative communities to supplement or replace the employment organization.

Many, however, may find that the home and the family will assume even more importance as the core community in their lives. More time in and around the home will mean that those who live there will fill more of our lives. It may not always be, however, the traditional nuclear family which lives in that home. A changed and enlarged role for the family may mean a changed and enlarged definition of the family. A new future for work is very possibly going to mean a new but different future for the family.

Contrary to the impression we may have formed, families are still with us and still fashionable. It is quite true that the number of people living on their own has doubled in the last twenty years. It is also quite true that many people live in houses with unrelated people (what *Social Trends* calls 'other households'), but in 1981 eight out of ten people in Britain still lived in a family unit headed by a married couple and a further 5 per cent in a one-parent family.[10] It is the 80/20 rule again.

But within that 80 per cent things are changing. It is not the same family unit all the time. One in three new marriages is ending in divorce, resulting in a growing tradition of serial monogamy and a new concept of the extended family, in which one has step-parents in place of uncles and aunts and step-brothers and -sisters as the new cousinry. It is a commentary on the new extended family that the latest Divorce Bill seeks to protect the rights of *grandparents* to continue seeing their natural grandchildren. Many of those who live with unrelated people in 'other households' might think of their household as a 'family' even if not a conventional one.

Nor do families continue to behave in the same way. Only 14 per cent of them, says the Study Commission on the Family, follow the traditional pattern of a male supporting a dependant wife and children. Married women, as we have seen, have increasingly joined the workforce (62 per cent at latest count),[11] although over half of those are in part-time work. It is still primarily the woman's job to run the home, whatever else she may do as well. Nor are the jobs she can get yet comparable with those of her husband or other men (see box 7.4). Will that change? It may. In fact, many would argue that it has to if we are going to share out *all* work more evenly, not just the glamorous (because it is paid and official) job work. It may change because other things are happening to families. Consider the following possibilities.

BOX 7.4   WOMAN'S PLACE AT WORK

In 1881 44 per cent of working women were in domestic service. In 1981 over half of working women were in 'intermediate and junior non-manual jobs', mostly clerical and secretarial, mostly low-status.

Things have not changed that much. Will they in future?

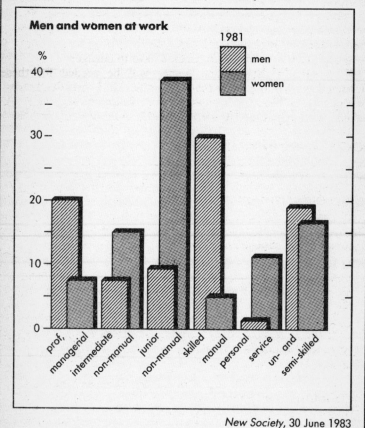

Men and women at work

1981

New Society, 30 June 1983

Families could become larger and looser

That does not mean that everyone will have more children, but rather that what we think of as our 'family' will include more people. Not only will divorce and remarriage bring more relatives into being, but increasingly it may also be more economic for more people to live together as a family unit. The 'household family' will grow. Hitherto progressive affluence has meant that households have shrunk progressively as young people have left home earlier and older people have stayed in their own homes.

Less paid work at the beginning and end of our working lives is likely to reverse that trend. Teenagers may not leave home, even if they want to, if they have only an apprentice's wage to live on. The grandparents of tomorrow may find that a pension no longer covers the cost of running and heating a separate establishment. Young people already share flats and houses because it is cheaper and more companionable. When a 'stable relationship' turns into marriage, is it so obvious that living arrangements have to change?

What sorts of houses and homes will be needed for these flexifamilies? Will the new extended families need extended homes, or will they be network families, linked by ever-cheaper telephone cables, visiting and talking with each other but not living together?

BOX 7.5   AMERICAN FAMILIES IN 1990

Husband-wife households with only one working spouse will account for only 14 per cent of all households as compared with 43 per cent in 1960.

Wives will contribute about 40 per cent of family income compared to about 25 per cent now.

At least thirteen separate types of household will eclipse the conventional family, including such categories as 'female head, widowed, with children' and 'male head, previously married, with children'.

More than a third of the couples first married in the 1970s will have divorced; more than a third of the children born in the 1970s will have spent part of their childhood living with a single family.

'The Nation's Families 1960–1990'
Joint Centre for Urban Studies of MIT and Harvard University
(summarized by John Naisbitt, *Megatrends* (Warren Books, New York, 1983)

Houses could be busier

The growth of the grey (domestic) economy will make the home more like a workplace than a motel for commuting workers. Less affluence and more time, particularly at the beginning and end of life, will make it sensible to do for oneself what we previously paid others to do for us. As self-employment grows, more people will be doing paid work from home for at least part of the time, or will be doing work at home that they used to buy in from outside.

Families may be more single-status

If both partners are around more, there will be a better chance to share all the roles in the family. The new gadgetry helps to make more jobs in and around the home more appealing to both sexes. Either partner can now cook or replace the plastic piping. Will each take shares in child-minding?

At the same time the tax and benefit system may well give up trying to tie allowances to family status but will instead offer benefits to the individual who needs them. It is the minder of the child who needs the child benefit allowance, irrespective of sex, and a married man's allowance is an anomaly in these times. Paradoxically, as Dr Eichler points out in Canada, 'in order to serve families as social units best, and to avoid discrimination against certain types of families, it is necessary to treat people *administratively* as individuals, so that they can live together socially in families.'[12]

Families will be held responsible

The education of children under 5, the housing of teenagers not yet at work, the care of the over 75s – should these be the responsibility of the family or of the state? The trend has been, gradually, for the state to take responsibility, but this trend has now come to an end because it is so expensive.

It could make a lot of sense to push these responsibilities back on to the family, both economically and sociologically. But the state must help – it would pay it to do so. Proper compensation would then allow both adults to spend time at home instead of having to work extra hours for extra pay. Money is an important part of it, but so is the kind of professional and emotional support that is required in these situations. Even if much of this were provided by self-help groups in the local community, these would in turn need to be provided with buildings and access to professional advice.

More time at home, less public money and more elderly relatives are strong inducements to return responsibility to the family. Let us hope it is done properly, not on the cheap or as a last resort.

The signs point to an extended role for the extended family or home. This would add a major 'caring' element to the portfolio of work, would confer new meaning on the lives of many of the givers and many of the receivers, would turn the home into a community rather than a hotel and would do much to balance out the lives of men and women.

But things could go the other way. Without a proper adjustment to our working patterns, which would make self-employment and the 50,000-hour job respectable and financially possible, without higher status for the work of the grey economy of the household, without proper assistance from the state for the care of dependants, the new role could become another burden on the backs of women, pulling them back from jobs, driving them into dependency once more and turning the home into a prison, not a community.

BOX 7.6   THE FAMILY TO THE RESCUE

The number of children under 5 was forecast to increase by 500,000 in the period between 1983 and 1986, yet the Conservative Government (in its 1983 White Paper on public spending) was planning for a *fall* in the participation rate of under-5s in nursery and primary schools. The family is expected to fill the gap.

The over-75s will grow by 90,000 between 1976 and 2001, and once again the Government expects that this will be taken care of by the 'community', which in the majority of cases means the family.

The choice is ours. To propel things in the right direction, we need to:

(1) give official status to domestic work and the care of dependants through the taxation and benefit system;

(2) move towards a guaranteed basic income system in order to encourage more people to spend more time at home working in the domestic economy and to reduce unnatural dependency;

(3) make it easier to turn homes into workplaces through better design and more relaxed rating regulations;

(4) recognize that the tasks of looking after oneself and caring for children and older parents are among the hardest, most worthwhile and most satisfying forms of work;

(5) recognize that divorce has replaced death as the commonest premature end to marriages and accept that there can be benefits in serial monogamy as part of a new concept of the family;

(6) educate men and women to do *all* the tasks in the home;

(7) provide more home-relief services and home advice, perhaps through state-supported voluntary agencies, to give those who are doing most of the caring some relaxation and some support;

(8) find ways of building homes which combine privacy, community and flexibility in order to meet the evolving and changing family boundaries.

And if we choose not to do these things, what then? Present trends will continue – trends towards a more and more individualist existence. In 1981 22 per cent of the population lived on their own (strikingly up on the 12 per cent who did so twenty years earlier).[13] In 1981 the nuclear family was down to two adults and 1.9 children and, in the case of 900,000 families, one adult and 1.6 children.[14] The economics of tiny households do not make much sense. The overheads to rent or mortgage, rates, heating and lighting are still there, with fewer to contribute to them. When times get hard these overheads consume a disproportionate amount of income, cutting into true disposable income and the freedom to enjoy society. When times get even harder the cost of the overheads becomes a charge on the state.

Socially, individual living leads to isolated living. Money, or the lack of it, combines with inclement weather to drive one indoors to one's own company or to a group too small to contain all the varieties of human humours. Privacy is important, but privacy overdone becomes loneliness. Lonely individuals and tiny families are not the best basis for bringing up children or caring for the old. They are, indeed, a recipe for more institutional care, more social welfare devices.

We have to face the facts. Financially, most people will be poorer, even if they are better off in other ways. The state will always be strapped for cash. The calls on its purse will always exceed the likely income from taxation. We will not be affluent enough, as individuals or as a society, to afford the kind of individualist living that is becoming the norm. If we do not want to become a withdrawn society, broke and too shy to leave our homes except for work, we need to welcome the idea of looser, larger families that today lie on the edges of society. We need to find ways to combine privacy with community through a variety of networks that allow us to belong and yet be alone. It is a time to choose.

BOX 7.7   WHICH WAY WILL IT GO?

The concept of service housing is increasingly popular in Sweden. Apartments are grouped around a central facility which provides a common child-care centre, a nursery for babies, and an after-school activity programme. There is also a common kitchen and dining-room where they can eat if they want to, although each apartment also has its own kitchenette. They usually also share laundry, gardening and cleaning services: privacy and community through a physical network organization.

Under the heading 'Small is Beautiful' David Crawford in the *Guardian* (19 November 1983) reports that over the next ten years the number of single-person households is expected to increase by 1.2 millions, representing 75 per cent of the total increase in new households in this period. Against this figure only 11 per cent of our houses are single-bedroomed homes, and he therefore reports that developers are now interested in this potentially expanding market.

We need, above all, to see opportunity in the facts when we face them. By specializing the functions of men and of women in the past we have actually deprived both sexes. Women have been articulate in describing how it feels to be excluded from the life that has traditionally been open to men. But men too have largely been prevented by this specialization from giving expression to the more feminine side of their personalities. All of us have bits of the masculine and bits of the feminine within us. The mix is not even, of course, and can vary from person to person, but to confine a man's expression of himself to the purely masculine is as inhibiting to him and to society as it is to a woman to be locked into the housekeeping and caring role. The portfolio flexilives that are likely to be forced upon us by the changing future of work may push us into discovering that we all have unused capacities within us. Most of us, it has been said, discover and use only a quarter of our talents before we die. Circumstances could force up that proportion.

Men, of course, have historically been very adept at turning any new situation to their advantage. Cooking as a domestic chore has traditionally been woman's work, but when it comes a profession then it is predominantly male chefs who do it. It would be sadly ironic if men, forced back into the domestic economy by the changing job situation, turned housework into a high-status occupation. We do not have to leave it to chance; we can choose, each of us individually. As far as the family is concerned, nobody has to wait for permission, even though some of the measures listed above could help to make things easier. The unspoken contracts within a family can be renegotiated at any time. The changing future of work will be as good a catalyst as any.

WHAT SORT OF SOCIETY?

At one level that is a simple question. Most people would like it to be rich, free and just or, as the French *révolutionnaires* had it, a combination of liberty, equality and fraternity. Unfortunately it is not as easy as that because the slogans manage to embrace ambiguity and incompatibility in one phrase. We cannot, for instance, be simultaneously free and equal, for if everyone were free to do what he or she liked, some would flourish and many would be trampled upon; equality would be a mirage. To the incompatibility of liberty and equality we have to add the ambiguity of justice. Is justice giving everyone what they *deserve*, by way of reward or punishment, or is it giving everyone what they *need*, or is it, as some would argue, giving everyone the same? These are hoary philosophical questions which have split political views for thousands of years, with those on the right favouring liberty and the justice of rewards and punishments, while those on the left have favoured more equality and the justice of needs or at least equality.

The questions are again going to be thrown into high relief because the changing future of work means that we shall have to rethink the shifting compromises that we have had to find between them. Should, for instance, those without jobs be treated well or badly? It depends upon what view of justice we hold. Does liberty mean the freedom to starve if you do not find work, or does equality insist that all must have some access to the goods and services of society? More wealth tends to make the compromises more palatable, but we are unlikely to be in that position for ten years at least. The compromises, therefore, will be difficult now. They will, in many cases, be decided upon ultimately by politicians, but politicians are elected by individuals and influenced by public opinion. These matters are too important to be delegated to politicians until we have each made up our own mind. Here the issues are sketched out very briefly to start the thinking.

Freedom or equality?

In the end there has to be a compromise, since absolute freedom and total equality are both impractical. The compromise can be an uneasy marriage of opposites. Britain's social policy, for instance, is a mixture of equality through *guaranteed rights* (to education and health care, for instance), and benefits based on *contribution*, (the freedom of the insurance principle). That should result in a bottom-floor equality and the freedom of 'benefits in return for contributions rather than free allowances from the state, which is what the people of Britain desire', as Beveridge said in his 1942 Report which laid the foundation of our current welfare state. This works well until the contributions prove inadequate to fund the kinds of benefits that are needed. Long periods of unemployment with no contributions, for example, result in inadequate (unequal) levels of unemployment benefit. Does freedom mean the unintended freedom to be poor?

The principle of liberty would, for instance, require people to take more responsibility for parts of their own pensions. It might go farther. New developments in medicine (orthopaedics, orthodontics and cancer screening) might well flourish outside the National Health Service, just as many parts of further education have reached beyond the remit of the education authorities. Liberty would argue that people should take responsibility for financing this for themselves. Equality would argue that this is unfair because the rich or the provident will always have an unfair advantage. But to provide everything for everyone (equality) would be impossibly expensive, with the result that equality has to be regulated (by assessment of need or by waiting lists). The insurance principle about which Beveridge was so adamant did, in fact, turn out to be very unfair to disabled people, to the unemployed young, to married women and to the poorly paid. But a health service that provided heart transplants on demand would be beyond our means, which means that someone (*not* the consumer) must decide what is within our means. Equality, in other words, tends to be expensive and authoritarian, while liberty is democratic but unfair. You cannot win!

What kind of justice?

The technicalities of retributive as opposed to distributive justice need not concern us here, but the principles do. A society which is anxious to reward its leaders, its entrepreneurs and its successful professionals and workers, to punish its criminals and cream off its young into select academies will be cheap to run but will rest on a heap of the neglected. On the other hand, a society which leans over backwards to offer help to those who need it will have to bleed its successful people in order to do that and will always feel that it is not doing enough. Should a teacher give more attention to the bright or to the stupid? It depends, in the end, on his or her definition of justice – deserts or need?

Once again compromise has its own morality. The usual consensus is that there has to be a bottom-floor of equality and some ceiling for liberty. The debate tends to be about the levels of the floor and the ceiling, whether they be taxation policy, health or education. The relevance of these concepts, and those of liberty and equality, to the future of work is immediate. For instance:

(1)   The cost of a national income scheme is likely to be so high that, in the name of equality, it will over-penalize the successful with high marginal rates of tax (unjust but equal).

(2)   On the other hand, a universal subsistence income, paid in cash, would increase individual choice and provide the freedom to work or not to work (liberty).

(3)   In which case wages, which would be in addition to the national income, would fall to market-clearing rates in some business areas, which would give everybody the chance to work (liberty), but would at the same time pull down other people's wages and reduce the tax base (unjust).

(4)   But to introduce a universal minimum wage to preserve income levels would imprison many in the grey (domestic) economy when they could be earning some money which would be good for them and the formal economy (equality but not liberty).

or:

(1)   Benefits should go only to those who used them, therefore we need a means test (justice of needs).

(2)   Yet this creates a 'poverty trap', which at the moment results in effective marginal tax rates of 60–90 per cent for people earning between £45 and £85 a week (no liberty).

(3)   But *universal* benefits (e.g. child benefit allowances or a national income scheme), while equal, end up being so inadequate that they do not meet the needs of those who really have to have them (equal but unjust in terms of needs).

The dilemma is perennial. No system can hope to satisfy fully the three high priests of liberty, equality and justice. Compromise requires that there be some of each. The individual will have to

BOX 7.8   ONE SET OF CRITERIA

In 1978 the Meade Committee on the Structure and Reform of Direct Taxation laid down six principles for a social security system:

(1) The system should provide an adequate minimum income for all, which . . .

(2) . . . should be provided without social stigma.

(3) The system should redistribute from the non-poor to those just above the poverty line once these first two criteria have been fulfilled.

(4) The benefits payable and the taxes necessary to finance them should be of such a form that discentive effects are minimized.

(5) The system should combine administrative simplicity with ease of understanding.

(6) The opportunities for abuse should be minimized.

Five years later these criteria remain unfulfilled — because they try to combine too much equality with too much liberty?

decide whether he or she wants an equal society with the danger that it may become increasingly authoritarian (allocating people to jobs perhaps or forbidding more than one wage-earner in a household) and hence a disincentive to the best talents, who may leave the country or give up. On the other hand, a society which maximizes liberty and incentives could end up with a proportion of well paid workaholics and a larger proportion of enforced idlers.

The dilemma has been put on this rather elevated, philosophical plane to encourage individuals to think beyond themselves and to look at society as a whole. A purely selfish attitude would lead those who feel deprived and unequal to vote for more equality and the justice of needs, while the more privileged would do the reverse. Such polarized voting makes constructive compromises more difficult, leaving society to lurch from liberty to equality and back again. The disappearance of jobs and the reorganization of work will make such a polarization of attitudes more likely than ever unless people look beyond themselves. We have to choose, and in considering this issue we are choosing not just for ourselves.

8

And in the End . . . ?

This is the last chapter. It is right that we should look at the big question – what will actually happen? It is also right to refuse to answer it, on the grounds that we are the arbiters of our own destiny and could refute, if we wanted to, anyone rash enough to prophesy. We can, however, lay out the alternatives and look at the possibilities.

The first alternative – full employment as we used to know it – is *not* going to work. Unless and until we abandon that notion we will pay little heed to anything in this book. Unfortunately, 'jobs for all' is a good slogan. It sounds worthy, and it allows its proponents to duck the new agenda. It links the right and the left of political parties, the right believing that an industrial revival will bring us back to full employment, and the left seeing the state as the employer of last resort. To both of them talk of alternative scenarios is defeatist, irresolute and immoral. For both of them, although by different means, the past can be recreated. This book, however, has argued that if 'jobs for all' means full-time, lifetime jobs for all who want them at good rates of pay, then full employment is not feasible in the foreseeable future.

Consider, once again, the following propositions.

(1) A more productive industry might create more wealth, but it will create more jobs only if output improves consistently faster than productivity. Given our need to remain internationally competitive and to jack up our productivity by at least 3 per cent p.a., this is an unlikely scenario.

(2) It would be unfair and unwise to burden productive industry and commerce with unproductive labour in an attempt to get them to solve our social problems.

(3) It was all the state could do in more affluent times to absorb the excess labour from industry. As industry and the tax base shrink, it will be even harder for the state to finance any further

growth in labour. Besides, what would they all do? The state requires, if anything, more teachers, social workers and nurses, not more unskilled labour.

(4)   Even if the state could pick up the slack, it would not cope with the million extra potential workers who we know will be joining the eligible workforce in the next fifteen years.

The full-employment equation of full-time, life-time jobs for all at good rates of pay is not viable. Something has to give. That is the first and the most crucial of the choices facing society. Is it better that all should have jobs even if many of them are very badly paid? Or is it better that all should have shorter or smaller jobs at good rates of pay? Or is it better that some should have no jobs at all, at least for a while? Each view has its advocates.

WHICH SCENARIO?

Tony Watts, in his examination of the future of work and education,[1] puts the choices quite starkly by proposing four alternative scenarios:

the unemployment scenario;
the leisure scenario;
the employment scenario;
the work scenario

They provide a useful way of focusing the options.

The unemployment scenario

Under this scenario we accept that unemployment is the necessary and inevitable price that we pay for bringing down inflation, for making industry and commerce more effective and internationally competitive and for keeping state expenditure within the boundaries of a shrinking tax base. Unemployment, we say to ourselves, is the lesser of two evils, the other one being international poverty.

We have been quite successful in conditioning ourselves to this view. As John Hughes has pointed out,[2] the standing army of the unemployed is now more than ten times the size of all our armed forces, yet, except in a very few areas, it is a largely unnoticed army. We have done away with the dole queue and have even convinced ourselves that many in this unnoticed army do not want work (the so-called 'scroungers'), do not need work (married women, older men) or do not deserve work (the unskilled and untrained). We comfort ourselves with anecdotes about the difficulty of finding a gardener or a window cleaner ('They can't be that hard-up, then') and about the new fashion for graduates on the dole.

It is a convenient scenario, for those in work. It helps if one can maintain that the condition is temporary, and if the invisible army remains invisible, particularly to those in power. The truth, unfortunately, is very different. As long as this scenario continues to praise legitimate and productive work and to reward it (average earnings consistently rise faster than the cost of living for those in work) and excludes a large section of the population, the results are predictable and well established. Unemployment is a scourge – particularly because, like a plague, it falls both on the just and the unjust: when a firm or an office closes the good worker as well as the bad is hurt. Most of all, as box 7.3 shows, it affects the loyal, committed, middle-aged working-class man.

The effect is particularly corrosive when unemployment hits whole communities, like Shotton in North Wales, Everton in Liverpool or Mablethorpe, all of which have more than 25 per cent of their workforce out of work. What the ripple effects of these sort of figures will be on the future generation can only be guessed at. To many it seems an unacceptably high price to pay for 'getting the economy right'.

The financial cost is also high – perhaps £12 billion in 1981, maybe £15 billion in 1983 – made up equally of benefit payments and tax that the unemployed would have paid had they been working.[3] It is a cost conveniently met, for the present, by North Sea oil, but even when that ceases the majority who are in work may feel it to be a price worth paying to preserve full employment for themselves.

The leisure scenario

This scenario seeks to turn employment to positive advantage. In its crudest form it suggests that a few, aided by a lot of machines, will do the work for the many, who will then be able to enjoy a life of leisure. Most people would be supported by the kind of national income scheme discussed in chapter 5, leaving work as an optional extra. People would, in effect, be paid to be consumers.

The optimistic version of this scenario sees a flourishing of the arts and of recreation, with enhanced community life, education devoted to the enrichment of self and others, a life devoid of toil. The pessimistic version turns to the science fiction of Kurt Vonnegut's novel *Player Piano* and Maureen Duffy's *Gor-Saga*, in which society is divided between those who work (and hold the power) and the masses who consume, living on the state.

In either form it seems hopelessly unrealistic. As James Robertson has pointed out, 'It just doesn't add up.'[4] The elite would have to have a high view of work, while the masses would have to believe that work was unimportant and that consumption was all that mattered. No society could combine such contradictory views of work, nor would those in work and in power relish the idea that they were doing it all so that the masses could enjoy not working.

The practical details are vague too. How would all this wealth be collected in order to be redistributed? Presumably not through income tax because there would be many fewer income earners. But any form of cash-flow tax or even profits tax would hit the competitiveness of the wealth-creating sector, and any form of high expenditure tax would be self-defeating, since it would have to be recycled immediately to the consuming masses to allow them to pay for the now more expensive goods.

More fundamentally, however, the idea of a leisured class is misconceived. It is true that the Greeks, the Romans, the Egyptians and even the English have had their leisured classes, but these were elites, and while they did no paid work, they all worked hard at governing, at the management of their estates, at patronizing the arts or at acting as benefactors to their communities. They seldom enjoyed real leisure: theirs was unpaid but glamorous work. When it did degenerate into pure consumption, as in Rome's last days, the civilization began to run down.

A leisured class at the bottom of society, however, rather than at the top, would have no glamorous but unpaid work to do. Bread and circuses, or their modern equivalents, dope and videos, have never proved to be the satisfactory basis for life or for society. On the other hand, a leisured class at the top of society, with the masses working for it, would be incompatible both with democracy (which would never grant so much to so few) and with technology (which needs the brains and skills of the elite to harness its possibilities). It is hard not to conclude that the leisure-class scenario is either the false dream of the poets among us or a well meaning way to whitewash inevitable unemployment.

Are there any signs that such a scenario is emerging? Opinions differ. The research on the long-term unemployed[5] consistently finds that they do not use their new free time to develop new interests or to take an active part in local affairs; instead they do more housework and watch more television. On the other hand, there are suggestions that there is a distinct youth culture in the areas of high unemployment which is based on consumption, amusement and 'unearned leisure', with no work ethic.[6] But maybe we expect too much of our young too soon. Among the professional middle classes some study and a lot of play and socializing are perfectly acceptable until the early twenties. Indeed, the smarter you are, the later you start real work. Should we expect the motivations of those who leave school at 16 to be different from those who leave educational institutions at 21? That there should be a non-work culture in late adolescence does not mean that a new leisure class has been born.

The idea of a separate leisure class may be misconceived, but that does not mean that there will not be much more leisure time in society as a whole. The question is, do we lump it all together and give it to one group of people, whether they want it or not, or do we spread it around? We *can* choose.

The employment scenario

This scenario is built on the premise that the only real form of work is a job, and harks back to that 23rd article in the Universal Declaration of Human Rights, that everyone has the right to 'free choice of employment'. Society has to make good that pledge.

A start would be to create more jobs by, for instance, infrastructure spending. Chapters 2 and 5 explored this route and its limitations. Although there are many good, even incontrovertible, arguments for going down it, the sums of money are so vast and the scale so daunting (see the example of a full-blown housing programme in chapter 5, p. 123) that in practical political terms no Government could afford more than a few sensible steps along the way. A public works programme would make a dent in the problem but would not solve it.

The next step might be to beef up the state service sector, even, some would say, to take over the voluntary sector on the basis that jobs are too precious to be given away for free. Why, for instance, encourage people to care for the young, for the infirm and the old for free and pay unemployment benefit to others who long for work? The answer is, first, economic – the tax base will not stand much more enlargement of the state sector. John Hughes has pointed out[7] that whereas in the first years of the 1970s there were 20 million people at work in the market sector, compared with 5 million in the public services or unemployed (a ratio of 4 to 1), the recent figures are 17 million to 8½ million (a ratio of 2 to 1). The second answer is that it is hard, perhaps wrong, to prevent people from giving their work for nothing in what is still a relatively open market place.

Others believe that we should positively encourage overmanning (or the employment of people in unnecessary jobs), as do the Japanese, for instance, who were estimated to have 2.5 million in employment, 'seat-warming', surplus to requirements.[8] If in 1983 it cost nearly £5,000 to maintain a married man with two children on the dole, it would not cost that much more to employ him somewhere, preferably usefully but, if not, at least to give him a workplace to go to. In a sense we already do this if we take account of the £20 billion a year which we spend to keep people employed in industries which would otherwise die. It would, according to this view, be cheaper to subsidize employment in a whole range of sectors rather than add to the unemployment benefit lists.

The arguments against this approach are that we would be in danger of (a) demeaning the concept of a job if it were seen as something for which society would not ordinarily pay and (b) creating a category of 'barely employables', who might in the end be *required* to take one of the artificial jobs rather than stay unemployed. It could, in other words, be a step towards the artificially full employment of totalitarian states, with the notions of directed labour which go with that full employment.

If the steps outlined above have been taken as far as they can be and the problem remains unsolved, the employment scenario turns to ways of sharing out work in the ways discussed in the final part of chapter 3. The underlying assumption is that x people doing 100,000 hours each is equivalent to perhaps $1\frac{1}{4}x$ people doing 50,000 hours, which leaves plenty of room for improved productivity and some room for increased rates per hour. Chapters 3 and 4 explored some of the ways in which this change might occur and concluded that it would not create enough jobs to solve the problem.

The argument of chapter 2 was that the process will be an irregular but inexorable progression. It will happen faster in some areas than in others. The professions, for instance, will use their monopoly arrangements to cling to the 100,000-hour job longer than most. Response to the change will be accepting rather than enthusiastic. Few will really want it because although the formula allows for increased hourly rates as well as increased productivity (which is why it will be permitted to happen), the total earnings from 50,000 hours will be much less than the total earnings from 100,000 hours. Furthermore, everyone will have to put away more of each wage packet, in one way or another, to pay for life beyond employment. It is a recipe for more leisure but also for less money in our pockets.

The work scenario

This scenario is the one broadly advocated by this book. In this scenario the job, and its earnings, are only part of work. The employment economy is only part of the whole economy, and money is only one of the rewards for work. There is, in this scenario, enough work for everyone, for work is priced at zero, as a gift, in many instances. There is enough potential status for everyone because all work carries some status. There could be enough money for everyone if we were all willing and able to top up our pay, our pensions or our social wages with part-time work.

To many people this scenario is just playing with words. Housework is not going to become more satisfying because we call it proper work rather than a tedious chore. They are right, in that the

BOX 8.1   IDEAS THAT HAVE CHANGED THE WORLD

When people look back on that tumultuous hundred years from 1880 to 1980 who, with the hindsight of a century or so, will seem the most powerful figures – Hitler, Stalin, Roosevelt and Churchill? Or Marx, Freud and Einstein?

The first four used political power to organize, and to kill, many millions. The other three were the authors of ideas that have changed the way in which whole societies, families and individuals think. They have cast their spell over our organizations, our law courts, our schools and our churches. In the end they may have had a more lasting impact than the first four put together, although none of them commanded more than a pen.

Marx brought to our attention the alienating aspects of organizations and institutions. Mankind cannot indefinitely be owned by others. If this form of ownership did not everywhere have the self-destructive force which Marx predicted, it was partly because people, both in management and in unions, took notice. The problems that come from replacing labour by capital run through this book, just as Marx foresaw, even if he was a century out in his dating.

Freud bequeathed to us the notion that we are not to be held totally responsible for our waywardness. Today we can, and do, blame a lot on our early life and social conditioning. A moral code which held us accountable for our actions, and on which a whole system of law and religion was founded, has had to be modified, for both good and ill.

Einstein may have set in motion the train of thought which led to the nuclear bomb, but he also triggered in more ordinary minds the idea that everything is relative and that even science has its element of randomness. That was to many a most releasing thought, encouraging more freedom of thought and expression, but it also undermined any idea of an absolute moral order.

new way of thinking has to lead to new rules, new taxes and new arrangements (the new agenda); without specifics it is just semantics. But they are wrong if they think that words cannot change things. There are few things more powerful than an idea whose time has come. If that idea changes the way people look at things, then it will ultimately change the way they organize things. It is a paradigm change. The most famous example is that of Copernicus, who by suggesting that the Sun, not the Earth, was the centre of the solar system changed nothing out there but ultimately changed the way we saw ourselves and the whole basis of science. 'Ultimately' is an important word, however, for paradigm changes take time to bite, but when they do they can be more powerful than their creators ever expected (see box 8.1).

HOW SHALL WE CHOOSE?

Two thoughts are immediately relevant: democracies believe in compromise, and the British embrace the inevitable.

We shall probably, in the end, muddle towards a compromise blend of all four scenarios. In one sense that will be a shame because the time may be what Beveridge called 'a revolutionary moment in the world's history', and such a moment should be 'a time for revolutions, not for patching'. On the other hand, democracies survive by compromise, by blending the interests of all parties, the 20 per cent and the 80 per cent. If they do not allow the needs and demands of the 20 per cent to infiltrate the 80 per cent they lay themselves open to overthrow as the 20 per cent becomes 30 per cent and then 40 per cent. The danger at present has to be the vested interests of the 80 per cent in that first scenario of 'requisite unemployment', which could, if it lasted long enough, create a new minority class in society, one largely excluded from society, one which ruled out compromise in favour of real revolution.

The British, however, are not a revolutionary people. This, I believe, is due largely to their ability to recognize the inevitable and to make it their own. The establishment has long used the sociological technique of 'co-opting' to disarm its opponents. Its institutions have proved sufficiently flexible over time to be enduring. Those that do not flex with the times are allowed to wither gracefully until their substitutes spring up in their places. It is interesting to observe how the British Empire changed into the British Commonwealth, and then, by almost unnoticed degrees, into the Commonwealth, as it adapted to new realities and new possibilities. Change by adoption and adaptation, by withering and rebirth, is, however, much slower and less dramatic than change by revolution. It is not until you look a long way back that you can see that things used to be different.

But no choice, even of compromise, is going to be free. Every choice will mean sacrifice of something by someone.

The sacrifices of the unemployment and leisure scenarios are plain to see. The brunt will be born by the unemployed or the new 'leisured class', although those remaining in work will have higher taxes to pay. In the employment and work scenarios the sacrifices are more evenly spread, in that everyone will have rather less employment and, inevitably rather less money. In these two scenarios, but particularly in the work scenario, those who have most (in the way of jobs, opportunity and the potential for wealth) will be called upon to sacrifice most. In the unemployment and leisure scenarios those who have least will sacrifice most.

It is clear which way compassion and the justice of needs will point. It is less clear which way the 80 per cent in a democracy will choose to go. Sacrifices after all, are willingly made only for great love or great causes. The French *révolutionnaires* were perhaps right in suggesting that liberty will make sacrifices for equality only in conditions of fraternity. But in a society which is already edging towards the unemployment scenario fraternity can be an illusion.

This, then, is the most crucial choice which the 80 per cent of the relatively fortunate in Britain have to make. Will they hang on to what they have and devil take the hindmost in a cruel world, or are they prepared to give up a part of their jobs, and with it a part of their income, for a fairer society, on the tenuous promise that it could be more exciting, fulfilling and rewarding for all if it happened? The pessimists, the fearful and the traditionalists would all say, 'No, Britain will not change but will continue to die slowly but, with luck, gracefully.' There are others, however, who like to think of Britain leading the world out of the industry-dominated era just as she led the world into it – unintentionally, haltingly and with many mistakes, it is true, but in the van all the same.

This book has been written in the belief that there is another way and that, were it offered to them in time, as a positive alternative, many of the 80 per cent might be prepared to follow it. Selfishness could be forced upon us by the lack of any positive alternative, for in a moral vacuum all one can do is look to oneself. Britons can be altruistic; the last stages of the British Empire showed that, as did the world wars, in horrific fashion. The creation of the welfare system after the war, along with the National Health Service and the Butler Education Act, were initial acts of altruism, even if they came to be thought of as entitlements.

Britain can be rich too, if no longer great. As chapter 5 argued, the British have traditionally excelled at many things, including: financial services, the arts, the theatre, the media (television and journalism), women's fashion, retailing, agriculture, consultancy, design engineering, medical research and architecture. All these successes have been characterized by small unit size, young people and young leaders, network management and the added value of brains of 'flair'. Britain's failure has been the management of large employment organizations. The future, therefore, is moving Britain's way if the traditional British attributes can be released by a change in the pattern of work, and if they can be applied to the innovative hardware which can be sold overseas as well as to the more exotic software of services. Whether her wealth will ever be enough to justify a position again among the great military powers is another matter. Great ambitions can be a great burden.

The British too, have always been admired by others, if not by themselves, for the integrity and competence of their Civil Service, their police and law courts, their care of the environment, the vitality of their theatre, music and artistic life, the stability of their political institutions, their health service and their universities. This is an infrastructure for a wealth of well-being which we would be foolish to squander, to damage or to take for granted. The portfolio lives that more and more people will live will need the security provided by sound administrative, judicial and health systems, variety in the arts and the natural environment and the enabling opportunities of education. Certainly there is more, much more, that needs to be done to improve our infrastructure, but the tradition is there for building on – Britain could be poised to enjoy her heritage more fully than for generations past as more people learn to enjoy that heritage.

The British, finally, are the rather forgetful inheritors of a strong religious tradition. The industrial revolution was enlightened and made more tolerable by Quaker and Methodist men and women of principle. It has been the British way to plant trees in all walks of life, trees which will outlive their founders. They have been able to think beyond themselves and beyond their generation. Is it too much to hope that their successors, the leaders of our institutions and organizations today, can also think as farsightedly?

Change in Britain, after all, comes seldom by edict from the centre but generally from a string of case law. Governments can frustrate or they can encourage and even codify what others have started. They can seldom start anything themselves. Case law that starts among the 20 per cent and spreads to the 80 per cent is what is needed. There are signs that the process has already started, that the fuse is lit. It should be a time for optimism.

AND FINALLY . . .

I am conscious not only of how lightly, even superficially, I have touched on problems of great complexity but also of all the topics which have not even been mentioned, yet could well deserve a chapter on their own – for example, the role of the developing world and our responsibilities to it, the problem of resources (particularly of energy), the arms industry and its economic consequences, the future of international trade and trading blocs. Each of these will have their impact on the future of work and on all our lives, but a short book imposes its own constraints, perhaps fortunately because in a longer book the problems would loom ever larger until finally they got in the way of any action.

The more I have pondered these questions, however, the more I have come to believe that we are imprisoned by our own assumptions. It is more important to challenge and to discard our assumptions than to explore all the possible tentacles of the future. We are fixated, both as a nation and as individuals, by the employment organization. Work is defined as employment. Money is distributed through employment. Status and identity stem from employment. We therefore hang on to employment as long as we can; we measure our success in terms of it; we expect great things from it, for the country and for ourselves; and we cannot conceive of a future without it. And yet, ironically, we are very bad at it because there is an individualist streak in all of us which agrees with Marx that it is alienating to sell ourselves or our time to another.

Break through that constraint and all sorts of things become possible, even if they are hard to visualize before they exist. The employment organization is the avenue of elms referred to at the very start of this book. Cut it down and a new landscape appears, a landscape yet to be designed and replanted. Leave the constraint alone and the scene will gradually decay and wither, ushering in a time of shabby gentility as Britain begins to resemble an old country house which has known better days and for which no one has yet found an alternative use, although the oil well in the backyard continues to provide for its dependants.

It is new imaginings which are needed to start a fresh debate among the 80 per cent. If we discard the shibboleth of employment, can we find a new liberty and a new energy without losing compassion and the requisite level of equality? The imaginings of this book may be wrong or unreal, but if they stir up further imaginings to help us to look beyond employment and beyond the status quo, they will have served their purpose.

References

1 A CHANGING WORLD

1 A good account of the development of economic thinking on unemployment is provided by Brian Showler in Brian Showler and Adrian Sinfield (eds.), *The Workless State* (Martin Robertson, Oxford, 1981).
2 Tom Stonier, *The Wealth of Information: A Profile of the Post-Industrial Economy* (Methuen, London, 1983), p. 15.
3 Emma Rothschild, 'Reagan and the Real America', in *New York Review of Books*, 5 February 1981, citing US Department of Labor, *Employment and Earnings 1980;* quoted in B. Jones, *Sleepers Wake!: Technology and the Future of Work* (Wheatsheaf Books, Brighton, 1982).
4 *Social Trends*, no. 14 (HMSO, London, 1983).
5 Ibid.
6 Ibid.

2 THE JOB SCENE

1 The figures in this analysis are all taken from *Social Trends* no. 14 (HMSO, London, 1983).
2 Stephen Humble, *Voluntary Action in the 1980s: Summary of the Findings of a National Survey* (The Volunteer Centre, Berkhamsted, 1982).
3 Richard Rose, *Getting by in Three Economies* (Centre for the Study of Public Policy, University of Strathclyde, 1983), pp. 29–30.
4 Tom Stonier, *The Wealth of Information: A Profile of the Post-Industrial Economy* (Methuen, London, 1983).
5 Marc Porat, *The Information Economy* (US Department of Commerce, Washington, 1977).
6 Barry Jones, *Sleepers Wake!: Technology and the Future of Work* (Wheatsheaf Books, Brighton, 1982), p. 48.
7 Stonier, *The Wealth of Information*.
8 Ibid.
9 Jones, *Sleepers Wake!*, p. 20.
10 Clive Jenkins and Barrie Sherman, *The Collapse of Work* (Eyre Methuen, London, 1979).

11 Jonathan Gershuny, *After Industrial Society?: The Emerging Self-Service Society* (Macmillan, London, 1979).
12 Stonier, *The Wealth of Information.*
13 *Report of the House of Lords Select Committee on Unemployment* (HMSO, London, 1981), pp. 69–72.

3 RETHINKING WORK

1 *The Economist*, 18 Sept. 1982, p. 69.
2 *General Household Survey 1980* (HMSO, London, 1982), pp. 85–6.
3 Richard Rose, *Getting by in Three Economies* (Centre for the Study of Public Policy, University of Strathclyde, 1983), p. 2.
4 A. Dilmot and C. N. Morris, 'What do we Know about the Black Economy?', *Fiscal Studies*, vol. 2, no. 1, 1981.
5 Rose, *Getting by in Three Economies.*
6 Gerald Mars, *Cheats at Work* (Allen and Unwin, London, 1982).
7 Helen Chappell, 'The Mauve Economy', *New Society*, 28 July 1983.
8 *Department of Employment Gazette*, June 1983.
9 *Economist*, 23 July 1983, p. 66.
10 Ibid., p. 69.
11 Rose, *Getting by in Three Economies*, p. 29.
12 Ibid., pp. 27–9.
13 Jonathan Gershuny, *After Industrial Society?: The Emerging Self-Service Society* (Macmillan, London, 1979).
14 Betty Friedan, *The Second Stage* (Michael Joseph, London, 1982), p. 42.
15 R. Kilpatrick and K. Trew, 'What Unemployed Men Do' (Department of Psychology, Queen's University, Belfast, 1982).
16 *Department of Employment Gazette*, April 1978.
17 M. White, *Case Studies of Shorter Working Time* (Policy Studies Institute, London, 1981).
18 *Working for a Future*, (Ecology Party, London, 1981), p. 36.
19 Giles Merritt, *World Out of Work* (Collins, London, 1982), p. 159.
20. *Social Trends*, no. 14 (HMSO, London, 1982).
21 *Department of Employment Gazette*, April 1978.
22 'Work-Sharing and Unemployment', IPM discussion document, May, 1983.
23 James Bellini, *Rule Britannica: A Progress Report for Domesday 1986*, (Cape, London, 1981).
24 Maureen Duffy, *Gor-Saga* (Methuen, London, 1981).

4 REORGANIZING WORK

1 Derek Sheane, 'Federalism', unpublished discussion paper (ICI, London, 1975).
2 Michael Cooley, 'Architect or Bee?' (CAITS, Polytechnic of North London, 1981).

3 John Naisbitt, *Megatrends* (Warner Books, New York, 1983).
4 F. Twaalfhoven and T. Hattori, 'The Supporting Role of Small Japanese Enterprises', Indivers Research, Schipol, Netherlands.
5 Ibid., p. 27.
6 Ibid., p. 11.
7 Ibid., p. 30.
8 Jenny Thornley, *Workers Co-operatives, Jobs and Dreams* (Heinemann Educational Books, London, 1982).

5 THE NEW AGENDA

1 Michael Pilch and Ben Carroll, 'State Pension Ages – Flexibility: the Key to Equality?', Lowndes Paper no. 1, (Lowndes Lambert Group Ltd, London, 1977).
2 Bill Jordan, *Automatic Poverty* (Routledge and Kegan Paul, London, 1981).
3 Tom Stonier, *The Wealth of Information: A Profile of the Post-Industrial Economy* (Methuen, London, 1983).
4 Keith Roberts, *Automation, Unemployment and the Redistribution of Income* (European Centre for Work and Society, 1982).
5 *Economist*, 17 September 1983, pp. 19–24.
6 Ibid.
7 Stonier, *The Wealth of Information*, p. 2.
8 *Social Trends*, no. 13 (HMSO, London, 1983).
9 Giles Merritt, *World Out of Work* (Collins, London, 1983), pp. 184 ff.
10 James Robertson, *The Sane Alternative*, rev. edn., (James Robertson, Ironbridge, 1983), p. 68.
11 Ivan Illich, *Disabling Professions* (Boyars, London, 1979).
12 Robertson, *The Sane Alternative*.
13 Figures provided by Volunteer, Washington DC, in 1983.
14 S. Hatch, 'Volunteering and Unemployment', Policy Studies Institute, London, 1983.
15 'Initiatives', *Journal of the Centre for Employment Initiatives*, November, 1983.

6 EDUCATING FOR TOMORROW

1 Randall Collins, *The Credential Society* (Academic Press, London, 1980).
2 Peter Wilby, 'Towards a Comprehensive Curriculum', in H. Pluckrose and P. Wilby (eds.), *The Condition of English Schooling* (Penguin, Harmondsworth, 1979).
3 A. H. Halsey, A. F. Heath and J. M. Ridge, *Origins and Destinations* (OUP, Oxford, 1980).
4 Barry Jones, *Sleepers Wake!* (Wheatsheaf Books, Brighton, 1982).
5 David Hargreaves, *Challenge for the Comprehensive School* (Routledge and Kegan Paul, London, 1982).

6 John White, *The Aims of Education Restated* (Routledge and Kegan Paul, London, 1982).

7 Auriol Stevens, *Clever Children in Comprehensive Schools* (Penguin, Harmondsworth, 1980).

8 'Experience, Reflection, Learning', Further Education Unit, Department of Education, 1978.

9 *The Practical Curriculum*, Schools Council Working Paper 70 (Methuen, London, 1980).

10 L. Katz, 'Skills of the Effective Administrator', *Harvard Business Review*, January 1955.

11 N. Evans, *The Knowledge Revolution* (Grant McIntyre, London, 1981).

12 Ibid.

7 CHOICES

1 Judith Humphries, *Part-Time Work* (Kogan Page, London 1983).

2 Ann McGoldrick, 'Early Retirement: A New Leisure Opportunity?' in 'Work and Leisure', Leisure Studies Association Conference Paper no. 15, Continuing Education Centre, Polytechnic of Central London, 1973.

3 Gurth Higgin, 'Scarcity, Abundance and Depletion: The Challenge to Management Education', Inaugural Lecture, Loughborough University of Technology, 1975.

4 Fred Hirsch, *The Social Limits to Growth* (Routledge and Kegan Paul, London, 1977).

5 Betty Friedan, *The Second State* (Michael Joseph, London, 1982).

6 For example, Warren Bennis, *The Unconscious Conspiracy* (Amacom, New York, 1974).

7 Quoted in T. J. Peters and R. H. Waterman Jr, *In Search of Excellence* (Harper and Row, New York, 1972).

8 Quoted by A. G. Watts, *Education, Unemployment and the Future of Work* (Open University, Milton Keynes, 1983).

9 P. Warr, 'Work, Jobs and Unemployment', *Bulletin of the British Psychological Society*, no. 36, 1983.

10 *Social Trends*, no. 13 (HMSO, London, 1983).

11 *General Household Survey 1980* (HMSO, London, 1982).

12 M. Eichler, *Families in Canada Today* (Gage, Ontario, 1983).

13 *Social Trends*, no. 13, 1983.

14 Ibid.

8 AND IN THE END . . .?

1 A. G. Watts, *Education, Unemployment and the Future of Work* (Open University, Milton Keynes, 1983).

2 J. Hughes 'Social Justice, Economic Crisis and the Welfare State', *Crisis and Theology*, William Temple Foundation, Occasional Paper no. 9, 1983.

3 A. Dilmot and C. Morris, 'The Exchequer Costs of Unemployment', *Fiscal Studies*, vol. 2, no. 3, November 1981.

4 J. Robertson, *The Sane Alternative*, rev. edn., (J. Robertson, Ironbridge, 1983).

5 E.g. 'Coping with Unemployment', Economist Intelligence Unit, London, 1982.

6 S. Frith, 'Dancing in the Streets', *Time Out*, 20 March 1981.

7 Hughes, 'Social Justice, Economic Crisis and the Welfare State'.

8 J. Sleigh and B. Boatwright, 'New Technology: The Japanese Approach', *Department of Employment Gazette*, vol. 87, no. 7, July 1979.

Further Reading

For those who want to delve further into any of the issues touched upon in this book there is a growing shelf of books to refer to. The ones listed below are those which I found most useful in my own studies.

Barry Jones, *Sleepers Wake! Technology and the Future of Work* (Wheatsheaf Books, Brighton, 1982)

This book, by an Australian politician, travels much the same route as *The Future of Work* but from an Australian point of view and in considerably more depth and detail. Australia turns out to have much the same dilemmas as Britain.

Giles Merritt, *World Out of Work* (Collins, London, 1982)

Giles Merritt, a journalist on London's *Financial Times,* took time off to explore the mounting unemployment policy in the industrialized world. His book ripples with pertinent statistics and examples but is pessimistic about any solution within the existing framework.

James Robertson, *The Sane Alternative,* revised edition (James Robertson, Ironbridge, 1983)

James Robertson, a well-known advocate of an alternative society, argues compellingly for a new approach to life and work, a breakthrough to a sane, humane and ecological world. He provides a long list of people who are doing things to promote this kind of world.

Tom Stonier, *The Wealth of Information: A Profile of the Post-Industrial Society* (Methuen, London, 1982)

Tom Stonier, best-known for his view that 10 per cent of the workforce could soon produce all our material needs, argues in this book that education and information could, if properly developed, make us wealthy as well as leisured.

Jonathan Gershuny, of Sussex University, has produced a number of recent books on the future shape of work in society, including the *New Service Economy* (Frances Pinter, London, 1983) on information technology, and *The New Household Economy* (OUP, Oxford, 1983) on the domestic economy, but his earlier and shorter book *After Industrial Society? The Emerging Self-Service Economy* (Macmillan, London, 1976) asks all the pertinent questions about the future of work.

Graeme Shankland, *A Guide to the Informal Economy* (Report for 'Work and Society', London, 1984)

This study, commissioned by the 'Work and Society' network, brings together in a very readable way all that is known about the various bits of the informal economy and those who work on it.

Tony Watts, *Education, Unemployment and the Future of Work* (Open University, Milton Keynes, 1983)

Tony Watts has nicely brought together the different scenarios for the future of work and the implications for education. His book is one sign that educationalists may be thinking farther ahead than many others.

John Naisbitt, *Megatrends* (Warner Books, New York, 1982)

For an American view of the way things are moving, read this account, written in a perceptive but anecdotal way. For a more visionary view, turn to:

Hazel Henderson, *Creating Alternative Futures: The End of Economics* (Berkley Windover, New York, 1978)

Bill Jordan, *Automatic Poverty* (Routledge and Kegan Paul, London, 1981)

There have been many accounts by economists of employment and unemployment but few attempt to look beyond the current structure to new alternatives. This one does, although not everyone might agree with Jordan's solutions.

Keith Roberts, *Automation, Unemployment and the Distribution of Income* (European Centre for Work and Society, Maastricht, 1982)

For those who want to examine further the workings of a National Income Scheme, this booklet examines in detail the ideal case.

Report for the Select Committee of the House of Lords on Unemployment (HMSO, London, 1982)

For a down-to-earth practical assessment of what is possible within the current structure you could hardly do better than read volume 1 of this report.

Guy Dauncey, *The Unemployment Handbook* (National Extension College, Cambridge, 1982)

This is a practical help-yourself guide for those who are already unemployed and cannot wait for society to reform itself.

Index